MEANING

MEANING

Michael Polanyi
and
Harry Prosch

THE UNIVERSITY OF CHICAGO PRESS
CHICAGO AND LONDON

THE UNIVERSITY OF CHICAGO PRESS, CHICAGO 60637
THE UNIVERSITY OF CHICAGO PRESS, LTD., LONDON

82 81 80 79   9 8 7 6 5 4 3

Library of Congress Cataloging in Publication Data

Polanyi, Michael, 1891–
   Meaning.

   Includes bibliographical references and index.
   1. Meaning (Philosophy)—Addresses—essays—
lectures. I. Prosch, Harry, 1917-   joint
author.   II. Title.
B105.M4P64   111.8'3   75-5067
ISBN 0-226-67294-8

To
CHARNER PERRY
*who first brought us
together*

CONTENTS

CONTENTS

PREFACE

In the spring of 1969 Michael Polanyi delivered a series of lectures at the University of Texas and at the University of Chicago entitled "Meaning: A Project by Michael Polanyi." I had been with him in Oxford, on a sabbatical leave in 1968-69, when he was preparing these lectures, and had read them with him before they were given. I found them very impressive, not only for the sequence of their development, but because they opened up vast new vistas in Polanyi's thought. The next year, 1970, he delivered another series of lectures at the University of Chicago. I was fortunate enough to be there with him, sharing with him a Willett Visiting Professorship in the Committee on Social Thought. A third series of lectures followed in 1971, at the University of Texas.

A number of lectures from these three series of lectures have been published as articles, but most of those that broke new ground in Polanyi's thought have not been published. These are the ones that spell out more particularly than any of his former works his explication of the sorts of meanings achieved in metaphors, poetry, art, ritual, myth, and religion. Most of the lectures that deal with these matters come from the series he delivered at Texas and Chicago in 1969.

In the spring of 1972 Professor Polanyi asked me if I could help him prepare these lectures for publication. His age then was such that he felt he needed help in bringing them out.

The core of the book, he said, should be based on the lectures that deal with the types of meaning mentioned above. These lectures should be introduced, he thought, by adaptations from some of his hitherto published articles. These would show the reader how the modern mind has destroyed meaning and how his own work on the reformation of epistemology and the philosophy of science has prepared the way for a possible restoration of meaning through the development of the notion of personal knowledge, as structured by the distinction between subsidiary and focal awareness.

The conclusion of the book should then show how the meanings established in science and those achieved in the humanities (those discussed in the main body of the book) can be brought into existential harmony through recognition of the existence of meaningful order in the world. This synthesis, together with the further recognition of the life of mutual authority, might then show the way toward a restoration of meaning in the life of contemporary man.

The general structure of the book, proceeding from the destruction of meaning to its restoration, was that of his Texas and Chicago lecture series of 1969.

I was both excited at the prospect of helping him show these ideas to the world and honored and pleased by his invitation. A leave of absence from Skidmore College and a Senior Fellowship from the National Endowment for the Humanities enabled me to accept his invitation and to begin work with him in England in February 1973.

A draft of the book was completed by me in October 1973. Professor Polanyi studied it until August 1974, when I returned to England once again to make a final decision with him about whether it should appear and, if so, in what form.

Substantively, therefore, this is Michael Polanyi's work. These are his ideas, expressed for the greatest part in his own language. In the work I have done on his lectures I have not consciously altered any of the ideas he has expressed in his numerous published and unpublished works. The reader will find in the Bibliographical Note at the end of this book a detailed listing of those works of Polanyi from which each chapter was adapted. I am largely responsible for the division of this work into its chapters; for the development of its continuity through the writing of various summary, supplementary, and bridging sections; and for the editorial work required—the final language, phrasing, and footnotes.

I should probably also explain here that wherever the pronoun "I" occurs in the text it is always Michael Polanyi who is speaking, relating a personal anecdote, feeling, or judgment which, by the nature of the case, could only be his.

I should like to express my deep gratitude to everyone who has been helpful in any way in making it possible for me to accomplish this work with Michael Polanyi. Most of all I thank Michael Polanyi himself for the opportunity extended to me to work in close association with a man whose breadth and depth of mind leave one with a sense of respect approaching awe and whose work must certainly be destined to leave an indelible mark upon the direction thought will take as it moves on toward the twenty-first century.

HARRY PROSCH

10 October 1974
Saratoga Springs, New York

ACKNOWLEDGMENTS

CHAPTERS 1 AND 13 HAVE BEEN ADAPTED IN PART FROM MICHAEL Polanyi's *The Logic of Liberty* (1951) and from his *Personal Knowledge* (1958), both published by The University of Chicago Press.

Chapters 2 and 3 have been adapted in part from Michael Polanyi's article "On the Modern Mind," *Encounter* 24 (May 1965): 12-20, and his "Logic and Psychology," *American Psychologist* 23 (January 1968): 27-43.

Chapter 2 was also adapted in part from Michael Polanyi's "The Study of Man," *Quest* (Bombay), No. 29 (April-June 1961): 26-35.

Chapter 12 was adapted in part from pages 63-79 of Michael Polanyi's *The Tacit Dimension* (Garden City: Doubleday and Company, Inc., 1966).

The remaining chapters, chapters 4-11, the central body of the text, are based on the series of lectures Michael Polanyi delivered at the University of Texas and at the University of Chicago in 1969, 1970, and 1971.

We wish to thank *Encounter*, *Quest*, *The American Psychologist*, The University of Chicago Press, and Doubleday and Company, Inc., for permitting us to make adaptive use of those of the above materials that were originally published by them.

Thanks should also go to Professor Richard Gelwick for the numerous discussions he found time to hold with Professor Polanyi relevant to the subject matter of this book during his stay in England and to John Brennan for his helpful criticisms of the text. Doris Prosch must also be thanked for the grueling hours she spent in proofing the text of the manuscript and for pointing out the many places where our initial choice of language would have required the wisdom of Solomon to disentangle our meaning; and to Magda Polanyi go our thanks for raising many practical questions about publication that mere philosophers would never think about and for, in general, holding things together.

We wish also to express our gratitude to the National Endowment for the Humanities for the award of a Senior Fellowship to Professor Prosch to enable him to work on this book in connection with Professor Polanyi in the spring of 1973. Thanks should also go to President Joseph C. Palamountain, Jr., and to Provost Edwin M. Moseley of Skidmore College for arranging a leave of absence for Professor Prosch, and to them and to the Skidmore College Faculty Research Committee for help in defraying expenses involved in Professor Prosch's further visit to England to finish the work with Michael Polanyi in the summer of 1974, as well as for help in defraying expenses involved in preparing this manuscript for publication.

MEANING

THE ECLIPSE OF THOUGHT

In a sense this book could be said to be about intellectual freedom. Yet its title, *Meaning*, is not really misleading, since, as we shall see, the achievement of meaning cannot properly be divorced from intellectual freedom.

Perhaps it could go without saying that intellectual freedom is threatened today from many directions. The ideologies of the left and the right of course have no use for it. In every one of these ideologies there is always some person, group, or party (in other words, some elite) which is supposed to know better than anyone else what is best for all of us; and it is assumed in these ideologies that it is the function of the rest of us—whether doctors, lawyers, or Indian chiefs—to support these "wise" decisions. The examples of fascism and of Marxist communism, especially as developed under Stalin, remain only too painfully present in the consciousness of twentieth-century man; moreover, the works of such writers as Milovan Djilas show us that even the most anti-Stalinist and liberal Communist regimes also engage in the repression of intellectual freedom.

We of the so-called Western world have opposed these totalitarian tyrannies—even to the extent of war. But we outselves have also threatened intellectual freedom. We have not, to be sure, drowned it in blood, as Hitler and Stalin did. Our threats have been much more devious. We have choked it with cotton, smothered it under various blankets. We have concealed our own affirmation of the value and freedom of our intellect under detached explanatory principles, like the pleasure-pain principle, the notion of the restoration of frustrated activity, the principle of conditioning—and even the concept of free choice itself! In such circuitous ways as these we have denigrated thought and all its works, demoting them to subordinate positions in which thought is conceived to function rightfully only when serving as a means to the satisfaction of supposedly more basic needs or wants, i.e., more material, more biological, more instinctive, more comforting.

Utilitarianism and pragmatism have both, in different ways, declared thought to possess a legitimate function or significance only in relation to social welfare—a welfare conceived largely in terms of physical and material satisfactions. The behaviorists, culminating in B. F. Skinner, have reduced thought to various forms of conditioned behavior and have directed us to look "beyond freedom and dignity"—beyond the life of self-control and self-direction—to the manipulated learning of a set of tricks supposed to be ultimately good for us to have learned. This learning would require us to be placed (by whom?) in a better-organized Skinner Box than that constituted by our present societies.[1] Old Protagoras, if we can trust Plato's interpretation of him, would have felt right at home with these ideas.

The only modern philosophic school that seems to exhibit respect for intellectual freedom is existentialism, but since it manages to smother the intellectual part of intellectual freedom under a more generic notion of freedom per se, it tends to weaken, in the end, our respect for intellectual freedom by reducing it in practice to the level of betting on the turn of a die. For these philosophers say there *are no grounds* for choices except the grounds we give ourselves, i.e., except the ones we choose. As Sartre puts it, value arises simply from our choices. What we choose, we value simply because we have chosen it (and apparently we remain scot-free at any moment to nonvalue it by simply un-choosing it). In other words, we do not choose (in his view) because we see the value of something. We see the value of something because we have chosen it. For him, therefore, every choice must ultimately be nonrational, because every rational choice, it is said, is ultimately grounded in a "prerational" choice. This position tells us, therefore, that there can be no reasons for our basic choices. Thought turns out to be of utilitarian value only—and then only when it *happens* to be of such value.

That this view may very well falter in its respect for intellectual freedom can be seen in the examples both Sartre and Simone de Beauvoir have given us by their on-again-off-again acceptance of various Communist suppressions of "bourgeois" artists and thinkers.

After all (as Sartre and de Beauvoir say—*sometimes*), no one governs innocently anyhow. *All* governments interfere with the exercise of *some* sorts of freedom. Since these philosophers (consistently) refuse to make any *philosophically* based value distinctions between different sorts of freedoms—or even between different uses of these different freedoms—they seem to echo old Bentham's remark: "Pushpin is as good as poetry." To repress one is no better and no worse than to repress the other.

We shall see, however, that the existentialists are closer to the truth in their view than any of the other academically popular Western philosophies, because there is a sense in which it is true that *determinative* reasons cannot be given for every choice—in fact, not for *any* choice. But the way existentialists have conceived this fact has generated unnecessarily antiintellectual attitudes, with disastrous consequences for the very freedom they value so fundamentally or, in their terms, "choose" so fundamentally.

It might be thought that our inquiry should now be directed to whether or not these erosions of respect for intellectual freedom in our day are justifiable. But even to raise this question is to answer it in the negative. For the attempt to judge any matter whatsoever is the attempt to think seriously about this matter, and such thinking cannot be undertaken without a tacit acceptance of the power of thought to reach valid conclusions. So our attempt to discover whether a right to intellectual freedom, i.e., the freedom to pursue subjects or problems intellectually, is or is not justified already assumes tacitly that it *is* justified.

Admitting, therefore, that the eclipse of our respect for freedom of thought cannot be justified, since it would require freedom of thought to justify it, we realize that nothing could have destroyed respect for freedom of thought but its own misuse; for it is only free thought that *could* call into serious question the validity of anything, including itself. Let us see therefore if we can discover how this self-destruction of thought came about.

From a careful study of the history of thought in our own time it is possible to see that freedom of thought destroyed itself when thought pursued to its ultimate conclusions a self-contradictory conception of its own freedom.

Modern thought in the widest sense emerged with the emancipation of the human mind from a mythological and magical interpretation of the universe. We know when this first happened, at what place, and by what method. We owe this act of liberation to Ionian philosophers who flourished in the sixth century B.C. and to other philosophers of Greece who continued their work in the succeeding thousand years. These ancient thinkers enjoyed much freedom of speculation but never raised decisively the issues of intellectual freedom.

The millennium of ancient philosophy was brought to a close by Saint Augustine. There followed the long rule of Christian theology and the Church of Rome over all departments of thought. The rule of ecclesiastic authority was impaired first in the twelfth century by a number of sporadic intellectual achievements. Then, as the Italian Renaissance blossomed out, the leading artists and thinkers of the time brought religion more and more into neglect. The Italian church itself seemed to yield to the new secular interests. Had the whole of Europe at that time been of the same mind as Italy, Renaissance humanism might have established freedom of thought everywhere, simply by default of opposition. Europe might have returned to—or, if you like, relapsed into—a liberalism resembling that of pre-Christian antiquity. Whatever may have followed after that, our present disasters would not have occurred.

However, there arose instead in a number of European countries—in Germany, Switzerland, Spain—a fervent religious revival, accompanied by a schism of the Christian church, which was to dominate people's minds for almost two centuries. The Catholic church sharply reaffirmed its authority over the whole intellectual sphere. The thoughts of men were moved, and politics were shaped, by the struggle between Protestantism and Catholicism, to which all contemporary issues contributed through their alliance with one side or the other.

By the beginning of the present century the wars between Catholics and Protestants had long ceased, yet the formulation of liberal thought still remained largely determined by the reaction of past generations against the old religious wars. Liberalism was motivated, to start with, by a detestation of religious fanaticism. It appealed to reason for a cessation of religious strife. This desire to curb religious violence was the prime motive of liberalism in both Anglo-American and Continental areas; yet from the beginning the reaction against religious fanaticism differed somewhat in these two areas, and this difference has since become increasingly accentuated, with the result that liberty has been upheld in the Western area up to this day but has suffered an eclipse in central and eastern Europe.

Anglo-American liberalism was first formulated by Milton and Locke. Their argument for freedom of thought was twofold. In its first part (for which we may cite the *Areopagitica*) freedom from authority is demanded so that truth may be discovered. The main inspiration for this movement came from the struggle of the rising natural sciences against the authority of Aristotle. Its program was to let everyone state his beliefs and to allow others to listen and form their own opinions; the ideas which would prevail in a free and open battle of wits would be as close an approximation to the truth as can be humanly achieved. We may call this the antiauthoritarian formula of liberty. Closely related to it is the second half of the argument for liberty, which is based on philosophic doubt. While its origins go back a long way (right to the philosophers of antiquity), this argument was first formulated as a political doctrine by Locke. It says simply that we can never be so sure of the truth in matters of religion as to warrant the imposition of our views on others. These two pleas for freedom of thought were put forward and accepted in England at a time when religious beliefs were unshaken and indeed dominant throughout the nation. The new tolerance aimed preeminently at the reconciliation of different denominations in the service of God. Atheists were refused tolerance by Locke on the ground that they were socially unreliable.

On the Continent the twofold doctrine of free thought—antiauthoritarianism and philosophic doubt—gained ascendance somewhat later than in England and moved straightway to a more extreme position. This position was first effectively formulated in the eighteenth century by the philosophy of Enlightenment, which was primarily an attack on religious authority, particularly that of the Catholic church. It professed a radical skepticism. The books of Voltaire and the French Encyclopedists, expounding this doctrine, were widely read in France, while abroad their ideas spread into Germany and far into eastern Europe. Frederick the Great and Catherine of Russia were among their correspondents and disciples. The type of Voltairean aristocrat, represented by the old Prince Bolkonski in *War and Peace*, was to be found at court and in feudal residences over many parts of Continental Europe at the close of the eighteenth century. The depth to which the *philosophes* had influenced political thought in their own country was to be revealed by the French Revolution.

Accordingly, the mood of the French Enlightenment, though often angry, was always supremely confident. Its followers promised mankind relief from all social ills. One of the central figures of the movement, the Baron d'Holbach, declared in 1770 that man is miserable simply because he is ignorant. His mind is so infected with prejudices that one might think him forever condemned to err. It is error, he held, that has evoked the religious fears which shrivel men up with fright or make them butcher each other for chimeras. "To errour must be attributed those inveterate hatreds, those barbarous persecutions, those numerous massacres, those dreadful tragedies, of which, under pretext of serving the interests of Heaven, the earth has been but too frequently made the theatre."[2]

This explanation of human miseries and the remedy promised for them continued to carry conviction with the intelligentsia of Europe long after the French Revolution. It remained an axiom among progressive people on the Continent that to achieve light and liberty you first had to break the power of the clergy and eliminate the influence of religious dogma. Battle after battle was fought in this campaign. Perhaps the fiercest engagement was the Dreyfus Affair at the close of the century, in which clericalism was finally defeated in France and was further weakened throughout Europe. It was at about this time that W. E. H. Lecky wrote: "All over Europe the priesthood are now associated with a policy of toryism, of reaction, or of obstruction. All over Europe the organs that represent dogmatic interests are in permanent opposition to the progressive tendencies around them, and are rapidly sinking into contempt."[3]

I well remember this triumphant sentiment. We looked back on earlier times as on a period of darkness, and with Lucretius we cried in horror: *Tantum religio potuit suadere malorum*—what evils religion has inspired! So we rejoiced at the superior knowledge of our age and its assured liberties. The promises of peace and freedom given to the world by the French Enlightenment had indeed been wonderfully fulfilled toward the end of the nineteenth century. You could travel all over Europe and America without a passport and settle down wherever you pleased. With the exception of Russia, you could, throughout Europe, print anything without prior censorship and could sharply oppose any government or creed with impunity. In

Germany—much criticized at the time for being authoritarian—biting caricatures of the emperor were published freely. Even in Russia, whose regime was the most oppressive, Marx's *Kapital* appeared in translation immediately after its first publication and received favorable reviews throughout the press. In the whole of Europe not more than a few hundred people were forced into political exile. Over the entire planet all men of European origins were living in free intellectual and personal communication. It is hardly surprising that the universal establishment of peace and tolerance through the victory of modern enlightenment was confidently expected at the turn of the century by a large majority of educated people on the Continent.

Thus we entered the twentieth century as on an age of infinite promise. Few people realized that we were walking into a minefield, though the mines had all been prepared and carefully laid in open daylight by well-known thinkers of our own time. Today we know how false our expectations were. We have all learned to trace the collapse of freedom in the twentieth century to the writings of certain philosophers, particularly Marx, Nietzsche, and their common ancestors, Fichte and Hegel. But the story has yet to be told how we came to welcome as liberators the philosophies that were to destroy liberty.

We have said that we consider the collapse of freedom in central and eastern Europe to be the outcome of an internal contradiction in the doctrine of liberty. But why did it destroy freedom in large parts of Continental Europe without producing similar effects, so far, in the Western or Anglo-American area of our civilization? Wherein lies this inconsistency?

The argument of doubt put forward by Locke in favor of tolerance says that we should admit all religions since it is impossible to demonstrate which one is true. This implies that we must not impose beliefs that are not demonstrable. Let us apply this doctrine to ethical principles. It follows that, unless ethical principles can be demonstrated with certainty, we should refrain from imposing them and should tolerate their total denial. But, of course, ethical principles cannot, in a strict sense, be demonstrated: you cannot prove the obligation to tell the truth, to uphold justice and mercy. It would follow therefore that a system of mendacity, lawlessness, and cruelty is to be accepted as an alternative to ethical principles and on equal terms. But a society in which unscrupulous propaganda, violence, and terror prevail offers no scope for tolerance. Here the inconsistency of a liberalism based on philosophic doubt becomes apparent: freedom of thought is destroyed by the extension of doubt to the field of traditional ideals, which includes the basis for freedom of thought.

The consummation of this destructive process was prevented in the Anglo-American region by an instinctive reluctance to pursue the accepted philosophic premises to their ultimate conclusions. One way of avoiding this was to pretend that ethical principles could actually be scientifically demonstrated. Locke himself started this train of thought by asserting that good and evil can be identified with pleasure and pain and by suggesting that all ideals of good behavior are merely maxims of prudence.

However, the utilitarian calculus cannot in fact demonstrate our commitment to ideals which demand serious sacrifices of us. A man's sincerity in professing his ideals is to be measured rather by the *lack* of prudence he shows in pursuing them. The utilitarian confirmation of unselfishness is not more than a pretense by which traditional ideals are made acceptable to a philosophically skeptical age. Camouflaged as long-term selfishness or "intelligent self-interest," the traditional ideals of man are protected from destruction by skepticism.

It would thus appear that the preservation of Western civilization up to this day within the Anglo-American tradition of liberty was due to this speculative restraint, which amounted to a veritable suspension of logic within British empiricist philosophy. It was enough to pay philosophic lip service to the supremacy of the pleasure principle. Ethical standards were not really replaced by new purposes; still less was there any inclination to abandon these standards in practice. The masses of the people and their leaders in public life could in fact disregard the accepted philosophy, both in deciding their personal conduct and in building up their political institutions. The whole sweeping advance of moral aspirations to which the Age of Reason opened the way—the English Revolution, the American Revolution, the French Revolution, the first liberation of slaves in the British Empire, the Factory Reforms, the founding of the League of Nations, Britain's stand against Hitler, the offering of Lend-Lease, U.N.R.R.A., and Marshall Plan aid, the sending of millions of food parcels by individual Americans to unknown beneficiaries in Europe—in all these decisive actions, public opinion was swayed by moral forces, by charity, by a desire for justice and a detestation of social evils, despite the fact that these moral forces had no true justification in the prevailing philosophy of the age. Utilitarianism and other allied materialistic formulations of traditional ideals remained merely verbal. Their philosophic rejection of universal moral standards led only to a sham replacement; or, to speak technically, it led to a "pseudo-substitution" of utilitarian purposes for moral principles.

The speculative and practical restraints which saved liberalism from self-destruction in the Anglo-American area were due in the first place to the distinctly religious character of this liberalism. As long as philosophic doubt was applied only to secure equal rights to all religions and was prohibited from demanding equal rights for irreligion, the same restraint would automatically apply in respect to moral beliefs. A skepticism kept on short leash for the sake of preserving religious beliefs would hardly become a menace to fundamental moral principles. A second restraint on skepticism, closely related to the first, lay in the establishment of democratic institutions at a time when religious beliefs were still strong. These institutions (for example, the American Constitution) gave effect to the moral principles which underlie a free society. The tradition of democracy embodied in these institutions proved strong enough to uphold in practice the moral standards of a free society against any critique that would question their validity.

Both of these protective restraints, however, were absent in those parts of Europe where liberalism was based on the French Enlightenment. This movement, being antireligious, imposed no restraint on skeptical speculations, nor were the standards of morality embodied there in democratic institutions. When a feudal society, dominated by religious authority, was attacked by a radical skepticism, a liberalism emerged which was protected by neither a religious nor a civic tradition from destruction by the philosophic skepticism to which it owed its origin.

Here, in brief, is what happened. From the middle of the eighteenth century, Continental thought faced up seriously to the fact that universal standards of reason could not be philosophically justified in the light of the skeptical attitude which had initiated the rationalist movement. The great philosophic tumult which started in the second half of the eighteenth century on the Continent of Europe and finally led up to the philosophic disasters of our own day represented an incessant preoccupation with the collapse of the philosophic foundations of rationalism. Universal standards of human behavior having fallen into philosophic disrepute, various substitutes were put forward in their place.

One such substitute standard was derived from the contemplation of individuality. The case for the uniqueness of the individual is set out as follows in the opening words of Rousseau's *Confessions*: "Myself alone.... There is no one who resembles me.... We shall see whether Nature was right in breaking the mould into which she had cast me." Individuality here challenged the world to judge it, if it could, by universal standards. Creative genius claimed to be the renewer of all values and therefore incommensurable. Extended to whole nations, this claim accorded each nation its unique set of values, which could not be criticized in the light of universal reason. A nation's only obligation was, like that of the unique individual, to realize its own powers. In following the call of its destiny, a nation must allow no other nation to stand in its way.

If you apply this claim for the supremacy of uniqueness—which we may call romanticism—to individual persons, you arrive at a general hostility to society, as exemplified in the anticonventional and almost extraterritorial attitude of the Continental *bohème*. If applied to nations, it results, on the contrary, in the conception of a unique national destiny, which claims the absolute allegiance of all its citizens. The national leader combines the advantages of both. He can stand entranced in the admiration of his own uniqueness while identifying his personal ambitions with the destiny of the nation lying at its feet.

Romanticism was a literary movement and a change of heart rather than a philosophy. Its counterpart in systematic thought was constructed by the Hegelian dialectic. Hegel took charge of Universal Reason, emaciated to a skeleton by its treatment at the hands of Kant, and clothed it with the warm flesh of history. Declared incompetent to judge historical action, reason was given the comfortable position of being immanent in history. An ideal situation: "Heads you lose, tails I win." Identified with the stronger battalions, reason became invincible—but unfortunately also redundant.

The next step was therefore, quite naturally, the complete disestablishment of reason. Marx and Engels decided to turn the Hegelian dialectic right way up. No longer should the tail pretend to wag the dog. The bigger battalions should be recognized as makers of history in their own right, with reason as a mere apologist to justify their conquests.

The story of this last development is well known. Marx reinterpreted history as the outcome of class conflicts, which arise from the need of adjusting "the relations of production" to "the forces of production." Expressed in ordinary language, this says that, as new technical equipment becomes available from time to time, it is necessary to change the order of property in favor of a new class; this change is invariably achieved by overthrowing the hitherto-favored class. Socialism, it was said, brings these violent changes to a close by establishing the classless society. From its first formulation in the *Communist Manifesto* this doctrine puts the "eternal truths, such as Freedom, Justice, etc."—which it mentions in these terms—in a very doubtful position. Since these ideas are supposed always to have been used only to soothe the conscience of the rulers and to bemuse the suspicions of the exploited, there is no clear place left for them in the classless society. Today it has become apparent that there is indeed nothing in the realm of ideas, from law and religion to poetry and science, from the rules of football to the composition of music, that cannot readily be interpreted by Marxists as a mere product of class interest.

Meanwhile the legacy of romantic nationalism, developing on parallel lines, was also gradually transposed into materialistic terms. Wagner and Walhalla no doubt affected Nazi imagery; Mussolini gloried in recalling imperial Rome. But the really effective idea of Hitler and Mussolini was their classification of nations into haves and have-nots on the model of Marxian class war. The actions of nations were in this view not determined, or capable of being judged, by right or wrong: the haves preached peace and the sacredness of international law, since the law sanctioned their holdings, but this code was unacceptable to virile have-not nations. The latter would rise and overthrow the degenerate capitalistic democracies, which had become the dupes of their own pacific ideology, originally intended only to bemuse the underdogs. And so the text of Fascist and National Socialist foreign policy ran on, exactly on the lines of a Marxism applied to class war between nations. Indeed, already by the opening of the twentieth century, influential German writers had fully refashioned the nationalism of Fichte and Hegel on the lines of a power-political interpretation of history. Romanticism had been brutalized and brutally romanticized until the product was as tough as Marx's own historic materialism.

We have here the final outcome of the Continental cycle of thought. The self-destruction of liberalism, which was kept in a state of suspended logic in the Anglo-American field of Western civilization, was here brought to its ultimate conclusion. The process of replacing moral ideals by philosophically less vulnerable objectives was carried out in all seriousness. This is not a mere pseudosubstitution but a *real* substitution of human appetites and human passions for reason and the ideals of man.

This brings us right up to the scene of the revolutions of the twentieth century. We can see now how the philosophies which guided these revolutions—and destroyed liberty wherever they prevailed—were originally justified by the antiauthoritarian and skeptical formulas of liberty. They were indeed antiauthoritarian and skeptical in the extreme. They even set man free from obligations toward truth and justice, reducing reason to its own caricature: to a mere rationalization of positions that were actually predetermined by desire and were held—or secured—by force alone. Such was the final measure of this liberation: man was to be recognized henceforth as maker and master, no longer as servant, of what before had been his ideals.

This liberation, however, destroyed the very foundations of liberty. If thought and reason are nothing in themselves, it is meaningless to demand that thought be set free. The boundless hopes which the Enlightenment of the eighteenth century attached to the overthrow of authority and to the pursuit of doubt were hopes attached to the release of reason. Its followers firmly believed—to use Jefferson's majestic vocabulary—in "truths that are self-evident," which would guard "life, liberty, and the pursuit of happiness" under governments "deriving their just powers from the consent of the governed." They relied on truths, which they trusted to be inscribed in the hearts of man, for establishing peace and freedom among men everywhere. The assumption of universal standards of reason was implicit in the hopes of the Enlightenment, and the philosophies that denied the existence of such standards denied therefore the foundation of all these hopes.

But it is not enough to show how a logical process, starting from an inadequate formulation of liberty, led to philosophic conclusions that contradicted liberty. We have yet to show that this contradiction was actually put into operation, that these conclusions were not merely entertained and believed to be true but were met by people prepared to act upon them. If ideas cause revolutions, they can do so only through people who will act upon them. If this account of the fall of liberty in Europe is to be satisfactory, it must show that there were people who actually transformed philosophic error into destructive human action.

Of such people we have ample documentary evidence among the intelligentsia of central and eastern Europe. They are the nihilists.

There is an interesting ambiguity in the connotations of the word "nihilism" which at first may seem confusing but actually turns out to be illuminating. As the title of Rauschning's book—*The Revolution of Nihilism*—shows, he interpreted the National Socialist upheaval as a revolution.[4] As against this, reports from central Europe often spoke of widespread nihilism, meaning a lack of public spirit, the apathy of people who believe in nothing. This curious duality of nihilism, which makes it a byword for both complete self-centeredness and violent revolutionary action, can be traced to its earliest origins. The word was popularized by Turgenev in his *Fathers and Sons*, written in 1862. His prototype of nihilism, the student Bazarov, is an extreme individualist without any interest in politics. Nor does the next similar figure of Russian literature, Dostoevski's Raskolnikov in *Crime and Punishment* (1865), show any political leanings. What Raskolnikov is trying to find out is why he should not murder an old woman if he wanted her money. Both Bazarov and Raskolnikov are experimenting privately with a life of total disbelief. But within a few years we see the nihilist transformed into a political conspirator. The terrorist organization of the Narodniki, or Populists, had come into being. Dostoevski portrayed the new type in his later novel *The Possessed*. The nihilist now appears as an ice-cold businesslike conspirator, closely prefiguring the ideal Bolshevik as I have seen him represented on the Moscow stage in the didactic plays of the early Stalinist period. Nor is the similarity accidental. The whole code of conspiratorial action—the cells, the secrecy, the discipline and ruthlessness—known today as the Communist method, was taken over by Lenin from the Populists. The proof of this can be found in articles published by him in 1901 and 1902.[5]

English and American people find it difficult to understand

nihilism, for most of the doctrines professed by nihilists have been current among themselves for some time without turning those who held them into nihilists. Great, solid Bentham would not have disagreed with any of the views expounded by Turgenev's prototype of nihilism, the student Bazarov. But while Bentham and other skeptically minded Englishmen may use such philosophies merely as a mistaken explanation of their own conduct—which in actual fact is determined by their traditional beliefs—the nihilist Bazarov and his kind take such philosophies seriously and try to live by their light.

The nihilist who tries to live without any beliefs, obligations, or restrictions stands at the first, the private, stage of nihilism. He is represented in Russia by the earlier type of intellectual described by Turgenev and the younger Dostoevski. In Germany we find nihilists of this kind growing up in large numbers under the influence of Nietzsche and Stirner; and later, between 1910 and 1930, we see emerging in direct line of their succession the great German Youth Movement, with its radical contempt for all existing social ties.

But the solitary nihilist is unstable. Starved for social responsibility, he is liable to be drawn into politics, provided he can find a movement based on nihilistic assumptions. Thus, when he turns to public affairs, he adopts a creed of political violence. The cafés of Munich, Berlin, Vienna, Prague, and Budapest, where writers, painters, lawyers, and doctors had spent so many hours in amusing speculation and gossip, thus became in 1918 the recruiting grounds for the "armed bohemians," whom Heiden in his book on Hitler describes as the agents of the European revolution.[6]   In much the same way, the Bloomsbury of the unbridled twenties unexpectedly turned out numerous disciplined Marxists around 1930.

The conversion of the nihilist from extreme individualism to the service of a fierce and narrow political creed is the turning point of the European revolution. The downfall of liberty in Europe consisted in a series of such individual conversions.

Their mechanism deserves closest attention. Take, first, conversion to Marxism. Historical—or dialectical—materialism had all the attractions of a second Enlightenment; taking off and carrying on from the first, antireligious, Enlightenment, it offered the same intense intellectual satisfaction. Those who accepted its guidance felt suddenly initiated into a knowledge of the real forces actuating men and operating in history, into a grasp of reality that had hitherto been hidden to them—and still remained hidden to the unenlightened—by a veil of deceit and self-deceit. Marx, and the whole materialistic movement of which he formed a part, had turned the world right side up before their eyes, revealing to them the true springs of human behavior.

Marxism also offered them a future of unbounded promise for humanity. It predicted that historic necessity would destroy an antiquated form of society and replace it by a new one, in which the existing miseries and injustices would be eliminated. Though this prospect was put forward as a purely scientific observation, it endowed those who accepted it with a feeling of overwhelming moral superiority. They acquired a sense of righteousness, and this in a paradoxical manner was fiercely intensified by the mechanical framework in which it was set. Their nihilism had prevented them from demanding justice in the name of justice or humanity in the name of humanity; these words were banned from their vocabulary, and their minds were closed to such concepts. But their moral aspirations, thus silenced and repressed, found an outlet in the scientific prediction of a perfect society. Here was set out a scientific utopia, relying for its fulfillment only on violence. Nihilists could accept, and would eagerly embrace, such a prophecy, which required from its disciples no other belief than a belief in the force of bodily appetites and yet at the same time satisfied their most extravagant moral hopes. Their sense of righteousness was thus reinforced by a calculated brutality born of scientific self-assurance. There emerged the modern fanatic, armored with impenetrable skepticism.

The power of Marxism over the mind is based here on a process exactly the inverse of Freudian sublimation. The moral needs of man, denied expression in terms of ideals, are injected into a system of naked power, to which they impart the force of blind moral passion. With some qualification the same thing is true of National Socialism's appeal to the mind of German youth. By offering them an interpretation of history in the materialistic terms of international class war, Hitler mobilized their sense of civic obligation which would not respond to humane ideals. It was a mistake to regard the Nazi as an untaught savage. His bestiality was carefully nurtured by speculations closely reflecting Marxian influence. His contempt for humanitarian ideals had a century of philosophic schooling behind it. The Nazi disbelieved in public morality the way we disbelieve in witchcraft. It is

not that he had never heard of it; he simply thought he had valid
grounds for asserting that such a thing cannot exist. If you told him
the contrary, he would think you peculiarly old-fashioned or simply
dishonest.

In such men the traditional forms for holding moral ideals had been
shattered and their moral passions diverted into the only channels
which a strictly mechanistic conception of man and society left open to
them. We may describe this as a process of *moral inversion*. The
morally inverted person has not merely performed a philosophic
substitution of material purposes for moral aims; he is acting with the
whole force of his homeless moral passions within a purely materialistic framework of purposes.

It remains only to describe the actual battlefield on which the
conflict that led to the downfall of liberty in Europe was fought out.
Let us approach the scene from the West. Toward the close of the First
World War, Europeans heard from across the Atlantic the voice of
Wilson appealing for a new Europe in terms of pure eighteenth-
century ideas. "What we seek," he summed up in his declaration of
the Fourth of July, 1918, "is the reign of law, based upon the consent
of the governed and sustained by the organized opinion of mankind."
When, a few months later, Wilson landed in Europe, a tide of
boundless hope swept through its lands. They were the old hopes of
the eighteenth and nineteenth centuries, only much brighter than
ever before.

Wilson's appeal and the response it evoked marked the high tide of
the original moral aspirations of the Enlightenment. This event
showed how, in spite of the philosophic difficulties which impaired
the foundations of overt moral assertions, such assertions could still
be vigorously made in the regions of Anglo-American influence.

But the great hopes spreading from the Atlantic seaboard were
contemptuously rejected by the nihilistic or morally inverted intelligentsia of central and eastern Europe. To Lenin, Wilson's language
was a huge joke; from Mussolini or Goebbels it might have evoked an
angry sneer. And the political theories which these men and their
small circle of followers were mooting at this time were soon to defeat
the appeal of Wilson and of democratic ideals in general. They were to
establish within roughly twenty years a comprehensive system of
totalitarian governments over Europe, with a good prospect of
subjecting the whole world to such government.

The sweeping success of Wilson's opponents was due to the greater appeal their ideas had for a considerable section of the populace in the central and eastern European nations. Admittedly, their final rise to power was achieved by violence, but not before they had gained sufficient support in every stratum of the population so that they could use violence effectively. Wilson's doctrines were first defeated by the superior convincing power of opposing philosophies, and it is this new and fiercer Enlightenment that has continued ever since to strike relentlessly at every humane and rational principle rooted in the soil of Europe.

The downfall of liberty which in every case followed the success of these attacks demonstrates in hard facts what we said before: that freedom of thought is rendered pointless and must disappear wherever reason and morality are deprived of their status as a force in their own right. When a judge in a court of law can no longer appeal to law and justice; when neither a witness, nor the newspapers, nor even a scientist reporting on his experiments can speak the truth as he knows it; when in public life there is no moral principle commanding respect; when the revelations of religion and of art are denied any substance; then there are no grounds left on which any individual may justly make a stand against the rulers of the day. Such is the simple logic of totalitarianism. A nihilistic regime will have to undertake the day-to-day direction of all activities which are otherwise guided by the intellectual and moral principles that nihilism declares empty and void. Principles must be replaced by the decrees of an all-embracing party line.

This is why modern totalitarianism, based on a purely materialistic conception of man, is of necessity more oppressive than an authoritarianism enforcing a spiritual creed, however rigid. Take the medieval church even at its worst. The authority of certain texts which it imposed remained fixed over long periods of time, and their interpretation was laid down in systems of theology and philosophy developed over more than a millennium, from Saint Paul to Aquinas. A good Catholic was not required to change his convictions and reverse his beliefs at frequent intervals in deference to the secret decisions of a handful of high officials. Moreover, since the authority of the church was spiritual, it recognized other independent principles outside its own. Though it imposed numerous regulations on individual conduct, many parts of life were left untouched, and these were governed by other authorities, rivals of the church, such as kings, noblemen, guilds, corporations. What is more, the power of all these was transcended by the growing force of law, and a great deal of speculative and artistic initiative was also allowed to pulsate freely through this many-sided system.

The unprecedented oppressiveness of modern totalitarianism has become widely recognized on the Continent today and has gone some way toward allaying the feud between the champions of liberty and the upholders of religion, which had been going on there since the beginning of the Enlightenment. Anticlericalism is not dead, but many who recognize transcendent obligations and are resolved to preserve a society built on the belief that such obligations are real have now discovered that they stand much closer to believers in the Bible and the Christian revelation than to the nihilist regimes, based on radical disbelief. History will perhaps record the Italian elections of April 1946 as the turning point. The defeat inflicted there on the Communists by a large Catholic majority was hailed with immense relief by defenders of liberty throughout the world, many of whom had been brought up under Voltaire's motto "Ecrasez l'infame!" and had in earlier days voiced all their hopes in that battle cry.

The instability of modern liberalism stands in curious contrast to the peacefully continued existence of intellectual freedom through a thousand years of antiquity. Why did the contradiction between liberty and skepticism never plunge the ancient world into a totalitarian revolution like that of the twentieth century?

We may answer that such a crisis did develop at least once, when a number of brilliant young men, whom Socrates had introduced to the pursuit of unfettered inquiry, blossomed out as leaders of the Thirty Tyrants. Men like Charmides and Critias were nihilists, consciously adopting a political philosophy of smash-and-grab which they derived from their Socratic education; and, as a reaction to this, Socrates was impeached and executed.

Yet whatever difficulties of this sort developed in the ancient world, they were never so fierce and far-reaching as the revolutions of the twentieth century. What was lacking in antiquity was the prophetic passion of Christian messianism. The ever-unquenched hunger and thirst after righteousness which our civilization carries in its blood as a heritage of Christianity does not allow us to settle down in the Stoic manner of antiquity. Modern thought is a mixture of Christian beliefs and Greek doubts. Christian beliefs and Greek doubts are logically incompatible; and if the conflict between the two has kept Western thought alive and creative beyond precedent, it has also made it unstable. Modern totalitarianism is a consummation of the conflict between religion and skepticism. It solves the conflict by embodying our heritage of moral passions in a framework of modern materialistic purposes. The conditions for such an outcome were not present in antiquity, when Christianity had not yet set alight new and vast moral hopes in the heart of mankind.

2

PERSONAL KNOWLEDGE

WE IN THE ANGLO-AMERICAN SPHERE HAVE SO FAR ESCAPED THE totalitarian nightmares of the right and left. But we are far from home safe. For we have done little, in our free intellectual endeavors, to uphold thought as an independent, self-governing force.

After the First World War our historians abandoned the vision of the Enlightenment that had evoked the dream of unlimited moral progress. Even before the war some academic movements of thought were leading in this direction. Positivism had set out to eliminate all metaphysical claims of knowledge. Behaviorism had started on the course that was to lead on to cybernetics, which claims to represent all human thought as the working of a machine. Sigmund Freud's revolution had started too, reducing man's moral principles to mere rationalizations of desires. Sociology had developed a program for explaining human affairs without making distinctions between good and evil. Our true convictions were being left without theoretical foundation.

One might indeed say that we too renounced the ideals of the nineteenth century. Alan Bullock wrote of Hitler that he was terrifyingly literal, and this was true also for Lenin. Our academic wisdom has lain, on the contrary, in *never meaning what we said*; our version of the disasters of Europe was a harmless whisper of their teachings.

History will not celebrate this performance, but it will still recognize that it kept us faithful at heart to the ideals of the nineteenth century.

Part of this chapter is adapted from Michael Polanyi, "Logic and Psychology," *The American Psychologist* 23 (January 1968): 30–34. Copyright 1968 by the American Psychological Association. Reprinted by permission.

Our mechanistic methods of inquiry have not killed altogether the generous feelings of our students. In the summer of 1964 hundreds of American students faced grim dangers by going to the aid of blacks in Mississippi. Nor have we, their teachers, entirely lost sight of the moral issues of our age. But our mechanistic methods have divorced our academic pursuits from these moral issues and made them merely "academic."

In general, therefore, our morally neutral account of all human affairs has caused our youth, and our educated people in general, to regard all moral professions as mere deceptions—or at best as self-deceptions. For once we induce ourselves to regard all established rules of moral conduct as mere conventions, we must come to suspect our own moral motives, and thus our best impulses are silenced and driven underground. Such self-suspicion does torment our age, and particularly many of our youth, seducing them into destructive forms of moral expression, since these alone seem proof against self-doubt. "I'm interested in anything about revolt, disorder, chaos, especially activity that has no meaning. It seems to me to be the road to freedom." Such a program (that of a popular songwriter) appeals to the conscience of our youth, for it is proof against suspicion of hypocrisy, whereas positive and constructive programs can all be suspected of being hypocritical. In other words, we also have been busily engaged in laying the groundwork for nihilism.

We have reason to think, however, that such effects of a false philosophy are not permanently debilitating, that, even when taught with all the compulsive powers of a totalitarian government, this philosophy may be unable to destroy the power of our moral convictions. Concluding his memoirs ten years after Stalin's death, the Soviet writer Ilya Ehrenburg spoke of "all the things that lie like a stone on the hearts of people of our generation." Such heavy hearts are part of a great movement, the movement of "revisionism," which seeks to redeem the ideals of the nineteenth century. It comes from Europe itself, where people have experienced to the full what it means to be subjected to a regime which denies reality to free thought and independent justice, which defines truth as party truth and reduces the arts to the service of propaganda. It is from these parts of Europe, mostly deeply lost to our ideals during the past fifty years, that a redemption of the ideas of the nineteenth century is once more dawning upon us.

At the 20th Congress of the Russian Communist Party, held in February 1956, Khrushchev first denounced Stalin's misdeeds in a secret speech. A few months later Polish and Hungarian writers were openly demanding freedom of thought. These men were leading Communist intellectuals who were recoiling from the theory that morality, justice, art, and truth itself were to be identified with the interest of the party. Hungarian Communist writers solemnly repudiated the teaching that political expedience can be a criterion of the truth and "after bitter mental struggles" vowed "that in no circumstances will we ever write lies." A few weeks later, the Hungarian people, led by these intellectuals, overthrew the Stalinist regime established by Rakosi.

This revolution, as well as that more recent ill-fated one in Czechoslovakia, was fought to gain recognition for the reality of intangible things: truth and justice and moral and artistic integrity. The Bolshevik attempt to establish an empire that denied this reality, though undertaken for high purposes and in the light of a sophisticated theory, had failed. It had proved unbearable. This passionate recognition of a metaphysical reality, irreducible to material elements, may well mark a turning point: it may serve as an axiom for any future political thought.

Writers in Poland, Hungary, and Czechoslovakia have been trying to find a place for the morally responsible individual within the Marxian conception of history. Early manuscripts of Marx, until recently unpublished, offer some substance for this, but the reviving of a few Hegelian ideas in the thought of the young Marx will not take us far. We need a theory of knowledge which shows up the fallacy of positivistic skepticism and supports the possibility of a knowledge of entities governed by higher principles.

Positivistic skepticism is one of a number of fallacies that have had their origin in modern science. In the days when it controlled all knowledge, religious dogma was a source of many errors. Now that the scientific outlook exercises predominant control over all knowledge, science has become the greatest single source of popular fallacies. This is not to denigrate science. Scientific genius has extended man's intellectual control over nature far beyond previous horizons. By secularizing man's moral passions, scientific rationalism has evoked a movement of reform which in the past hundred and fifty years has improved almost every human relationship, both public and private. A rationalist concern for welfare and for an educated and responsible citizenship has created an active mutual concern among millions of previously submerged and isolated individuals. Scientific rationalism has indeed been the main guide to intellectual, moral, and social progress since the idea of progress first gained popular acceptance about a hundred and fifty years ago.

Unfortunately, the *ideal* goals of science are nonsensical. Current biology is based on the assumption that you can explain the processes of life in terms of physics and chemistry; and, of course, physics and chemistry are both to be represented ultimately in terms of the forces acting between atomic particles. So all life, all human beings, and all works of man, including Shakespeare's sonnets and Kant's *Critique of Pure Reason*, are also to be so represented. The ideal of science remains what it was in the time of Laplace: to replace all human knowledge by a complete knowledge of atoms in motion. In spite of much that is said to the contrary, quantum mechanics makes no difference in this respect. A quantum-mechanical theory of the universe is just as empty of meaning as a Laplacean mechanical theory.

It is simply this sort of mechanical reductionism that is the heart of the matter. It is this that is the origin of the whole system of scientific obscurantism under which we are suffering today. This is the cause of our corruption of the conception of man, reducing him either to an insentient automaton or to a bundle of appetites. This is why science denies us the possibility of acknowledging personal responsibility. This is why science can be invoked so easily in support of totalitarian violence, why science has become the greatest source of dangerous fallacies today.

Let us look at some examples of the absurdities imposed by the modern scientific outlook. Listen to three authoritative voices denying the existence of human consciousness: (1) "... the existence of something called consciousness is a venerable *hypothesis*: not a datum, not directly observable ..."; (2) "... although we cannot get along without the concept of consciousness, actually there is no such thing"; (3) "The knower as an entity is an unnecessary postulate." These three statements were made, respectively, by Hebb, Kubie, and Lashley at a symposium on brain mechanisms and consciousness in 1954.[1] It is not that these distinguished scientists really believe that consciousness does not exist. They know, for example, that pain exists. But they feel obliged to deny the existence of consciousness, for it eludes explanation in terms of science.

You meet the same situation in the study of society. Anthropologists must try to describe social groups in strictly scientific terms. Most anthropologists thus insist on carrying out their analyses of society without mention of good and evil. A distinguished anthropologist has represented the unspeakably cruel murder of supposed witches as a cultural achievement. "Some social systems," he writes, "are much more efficient than others in directing aggression into oblique or non-disruptive channels. But there is no doubt that witchcraft is Navaho culture's principal answer to the problem that every society faces: how to satisfy hate and still keep the core of society solid."[2] Another anthropologist has described head-hunting as fulfilling an essential function in the societies in which it is practiced. "The religion of the Eddystone Islanders," writes Gordon Childe, "provided a motive for living and kept an economic system functioning." Eddystone culture proved wrong, in his view, only because head-hunting, by keeping the population down, made technical progress superfluous and so left the islanders a prey to British conquerors.[3]

For this kind of scientific anthropology social stability is the only accepted value and becomes, therefore, the supreme social value. Yet all the time we know, and the anthropologists know it as well as anybody, that the stability of evil is the worst of evil. They ignore this vital fact only for the sake of scientific detachment. The more absurd their attitude, the more it adds to their reputation for scientific rigor.

How difficult it is for this sort of detached "objectivity" to see a moral struggle for what it is is clear in the following incident. Professor Pipes tells us that when he wrote an essay on the Russian intelligentsia, in 1960, for *Daedalus*, he wanted to conclude it with a brief statement to the effect that the modern Russian intellectual has a very special mission to fulfill: "to fight for the truth," but that on the advice of friends he omitted the passage because it sounded "naive" and "unscientific." Four years later he decided it was *not* naive and unscientific to attribute to Russian intellectuals a "craving for truth." Yet even then he could define a "right to truth" (which quite obviously these men—and the Hungarians and the Poles and the Czechs—actually wish to have) only as the "right to surrender to one's impressions without being compelled for some extraneous reasons to interpret and distort them."[4]

We see that it became necessary for Professor Pipes to create a labyrinth of subterfuges in order to say ("scientifically") that a group of persons regards the seeking of the truth to be their right and duty; we also see that the involuted words he substituted for this single-minded commitment do not begin to express what is actually taking place in eastern Europe.

The roots of this perversion go deep. The rebellion of scientific rationalism against religious authority was based on the appeal to facts against dogma. Positivism merely pursued this movement to its logical conclusions by repudiating metaphysics along with dogma. The Viennese school of philosophy carried out this program by rejecting as metaphysical any statements about the world that are not verifiable in experience or that are not—according to Karl Popper's amendment—falsifiable by experience. This view in effect discredits all ethical statements. For if you say that it is wrong to bear false witness, you say something that cannot be proved or disproved by experiential facts. No conceivable occurrence, no measurement or observation, can decide whether any action is moral or immoral, just or unjust, good or evil. Hence, in this positivistic view of empiricism, to call something immoral, unjust, or evil is to speak with no empirical meaning; and it appears doubtful then whether such a statement could have any meaning beyond the kind of exclamation one may make when biting into a worm in an apple or when shouting to stop others from doing things one finds distressing.

Admittedly, this conception of moral judgment is felt to be unsatisfactory; for whenever we utter moral condemnation or approval, or else seek guidance in a moral dilemma, we always refer to moral standards assumed to be generally valid, and we revere men, like Socrates or Gandhi, who face death to uphold such standards. Hence the descendants of the Positivists are now engaged in great efforts to save moral standards from being cast out as altogether unfounded. But their efforts are in vain. As long as science remains the ideal of knowledge, and detachment the ideal of science, ethics cannot be secured from complete destruction by skeptical doubt.

In earlier ages philosophers could keep their gravest doubts among themselves. Hume brushed doubt aside for a game of backgammon. His great successor Kant still thought that there was no danger that philosophic skepticism would ever gain popular influence. Philosophy could never have much effect upon the masses. But ours is an age of philosophic mass movements. A glance at current books or at the daily newspaper reveals the same corrosive passion for destroying man's moral image of himself. Two little English books were written simultaneously in 1942, one entitled *The Abolition of Man*, the other *The Annihilation of Man*; the first was by C. S. Lewis, the second by Leslie Paul. C. S. Lewis took a schoolbook on English as a specimen of the debunking of moral and aesthetic sentiments by the teachings of our time. Lewis, after acknowledging that this debunking had started after the First World War with the laudable purpose of saving men from being swayed by nationalist propaganda, warned that the dishonoring of traditional ideals would merely divert man's moral passions into baser channels.[5] Leslie Paul's book bore out this view. He quoted the lines by which Oswald Spengler acclaimed Hitler in 1934: "Man is a beast of prey . . .; would-be moralists . . . are only beasts of prey with their teeth broken . . .; remember the larger beasts of prey are *noble* creatures . . . and without the hypocrisy of human morals due to weakness."[6] Observe the argument: (1) man is a beast, (2) his morals are hypocritical, (3) hypocrisy is revolting. Thus beastliness alone is honest and noble. Many seem to think this moral approval of brutality is only a German vice, but Simone de Beauvoir hails the glorification of crime and lust by the Marquis de Sade as great moral pronouncements and then identifies these teachings of crime and lust with the exposure of bourgeois ideologies by historical materialism. So the French Marxist writer transmutes bestiality into moral rebellion even as the Nazi historian does. Here again we see a moral inversion in which skepticism drives men's moral sentiments underground, whence they emerge, combined with sadism, as a creed of salvation by violence. Fascism thus converted patriotism into a cult of brutality, even as Marx converted utopianism into a science. Our age is racked by the fanaticism of unbelievers.

This is what C. S. Lewis meant by saying that science is the greatest source of dangerous fallacies today. The question is: Can we get rid of all these malignant excrescences of the scientific outlook without jettisoning the benefits which it can still yield to us both mentally and materially?

This appears to be a large order. But we can start mending this supposed break between science and our understanding of ourselves as sentient and responsible beings by straightening out our conception of scientific knowledge. Let us therefore do something quite radical, something quite forbidden by our current views of science. Let us incorporate into our conception of scientific knowledge *the part which we ourselves necessarily contribute* in shaping such knowledge. Let us proceed with a critique of the exact sciences in order to displace quite generally the current ideal of detached observation by a conception of *personal* knowledge.

Laplace's ideal of embodying all knowledge of the universe in an exact topography of all its atoms remains at the heart of the fallacies flowing from science today. Laplace affirmed that if we knew at one moment of time the exact positions and velocities of every particle of matter in the universe, as well as the forces acting between the particles, we could compute the positions and velocities of the same particles at any other date, whether past or future. To a mind thus equipped, all things to come and all things past would stand equally revealed. Such is the complete knowledge of the universe as conceived by Laplace.

This ideal of universal knowledge would have to be transposed into quantum-mechanical terms today, but this is immaterial. The real fault in the kind of universal knowledge defined by Laplace is that it would tell us absolutely nothing that we are interested in. Take any question to which you want to know the answer. For example, having planted some primroses today, you would like to know whether they will bear blossoms next spring. This question is not answered by a list of atomic positions and velocities at some future moment on May 1 of next year. *Primroses*, as such, are lost in the topography of *all* the atoms. Your question can be answered only in terms of primroses. The universal mind is utterly useless for this purpose unless it can go beyond predicting *atomic* data and tell us whether they imply the future blossoming or failure to blossom of the *primroses* planted today.

But let us shelve for the moment the question whether we can or cannot infer something about primroses, or about anything else we may be interested in, from a topography of atomic positions and velocities. It is enough to realize in the first place that, as it stands, Laplace's representation of the universe ignores all our normal experience and can answer no questions about it. We shall show that this shortcoming of the Laplacean scheme is due to a misunderstanding of the very nature of experimental science.

Consider the use of geographical maps. A map represents a part of the earth's surface in the same sense in which experimental science represents a much greater variety of experience. To use a map to find our way, we must be able to do three things. First we must identify our actual position in the landscape with a point on the map, then we must find on the map an itinerary toward our destination, and, finally, we must identify this itinerary by various landmarks in the landscape around us. Thus map-reading depends on the tacit knowledge and skill of the person using the map. Successful identification of actual locations with points on a map depends upon the good judgment of a skilled map-reader. No map can read itself. Neither can the most explicit possible treatise on map-reading read a map.

Turn now to the exact sciences, which Laplace had in mind when he was defining universal knowledge. The map is replaced here by formulas like the laws of planetary motion. These too are applied in three stages. First we make some measurements which yield a set of numbers representing our experience at the start; from these numbers we then compute, by the aid of our formulas, a future event; finally, we look out for the experience predicted by our computation. At both the beginning and the end we identify numbers with observed events, and this too is a kind of map-reading, for which we must rely once more on our personal skill. Numbers do not of themselves point to events.

People miss this point when they speak of the exact predictions made by the mathematical sciences. Take, for example, astronomy, which was very much in the mind of Laplace when he formulated his ideal of universal knowledge. You might think that Newton's laws could predict the exact position of the planets at any future moment of time. But this they can never do. Astronomers can merely compute from one set of numbers, which they identify with the position of a planet at a particular time, another set of numbers, which will represent its position at a future moment of time. But no formulas can foretell the actual readings on our instruments. These readings will rarely, if ever, coincide with the predicted numbers as computed from Newton's laws, and there is no rule—*and can be no rule*—on which we can rely for deciding whether the discrepancies between theory and observation should be shrugged aside as observational errors or be recognized, on the contrary, as actual deviations from the theory. The assessment in each case is a personal judgment.

Even the most modern instruments are affected by this uncertainty. There is ample evidence that even by using highly automated recorders we cannot exclude a personal bias that might affect a series of readings. Even the most exact sciences must therefore rely on our personal confidence that we possess some degree of personal skill and personal judgment for establishing a valid correspondence with—or a real deviation from—the facts of experience.

We may conclude quite generally that no science can predict observed facts except by relying with confidence upon an art: the art of establishing by the trained delicacy of eye, ear, and touch a correspondence between the explicit predictions of science and the actual experience of our senses to which these predictions shall apply.

You may feel that we are attributing undue significance to a small and perhaps altogether negligible coefficient in the structure of science, but this would be like excusing the housemaid's illegitimate baby on the ground that it is, after all, only a small baby. It is the principle that matters; and in fact the slight gap between theory and instrument readings turns out to be thin only in the way the edge of a wedge is thin—a wedge that will prove thick enough at its base to completely separate "knowledge" from "detached objectivity." Personal, tacit assessments and evaluations, we see, are required at every step in the acquisition of knowledge—even "scientific" knowledge.

Look at the buildings that compose the medical school of a modern university. You see row upon row of laboratories and dissection rooms, and you see a whole array of teaching hospitals. Students of chemistry, biology, and medicine spend a good half of their time in these places, where they seek to bridge the gap between the printed text of their books and the facts of experience. They are training their eyes, their ears, and their sense of touch to recognize *the things* to which their textbooks and theories refer. But they are not doing so by studying further textbooks. They are acquiring the skills for testing by their own bodily senses the objects of which their textbooks speak. Here there can no longer be any question of shrugging aside as a marginal factor the purely *personal* judgments by which the theoretical body of science is brought to bear on experience. Textbooks of chemistry, biology, and medicine are so much empty talk in the absence of personal, tacit knowledge of their subject matter. The excellence of a distinguished medical consultant or surgeon is due not to his more diligent reading of textbooks but to his skill as a diagnostician and healer—a personal skill acquired through practical experience. His professional distinction therefore lies in a massive body of personal knowledge.

Remember also that the fundamental concepts of the biological sciences are drawn from everyday experience, in which exact measurement plays no part. The existence of animals was not discovered by zoologists, much less by atomic physicists or chemists; nor was the existence of plants discovered by botanists. We learn to distinguish living beings from inanimate matter long before we study biology, and, when we do study it, we continue to use our original conception of life. Psychologists must know from ordinary experience what intelligence is before they can devise tests for measuring it scientifically. It is ordinary people, knowing the sufferings of sickness and the joy of recovery, who set medical science its task.

It is true that the progress of science is constantly molding and modifying our everyday conceptions. But when this is allowed for, it still remains true that there is a vast range of everyday knowledge, conveying delicate and complex conceptions, that serves as a guide to biology, medicine, psychology, and to the manifold disciplines that study man and society. And this knowledge is transmitted by adults to children as they grow up, in the form of practical arts, in the very same way that students are taught scientific skills and expert knowledge at the bedside and in the laboratory.

All this brings out squarely the general principle which limits the scope of the exact sciences, of which the Laplacean vision is the extreme idealization. Most of the questions in which we are interested are of the same kind as our question about the future blossoming of newly planted primroses. Answers to such questions must be given in terms of personal knowledge available to the layman, as corrected and expanded by the sciences—which rely in their turn on the further personal knowledge of experts. Laplacean predictions would convey none of this personal knowledge. To claim that a worldwide topography of atoms represents universal knowledge is to contradict the very principle of identification which must be used even in a mathematical theory if it is to bear upon experience. Hence, if the Laplacean vision or a similar ideal of the exact sciences succeeded in establishing itself as the total of man's knowledge, it would impose complete ignorance on us.

We must therefore amend our ideal of science by accrediting skills and connoisseurship as valid, indispensable, and definitive forms of knowledge. This amendment, we shall see, will open the way to a far-reaching relaxation of the tension between science and the nonscientific concerns of man. Let us observe how this is so by inquiring into the essential structure of knowing as an art.

A striking feature of knowing a skill is the presence of *two different kinds of awareness* of the things that we are skillfully handling. When I use a hammer to drive a nail, I attend to both, but quite differently. I *watch* the effects of my strokes on the nail as I wield the hammer. I do not feel that its handle has struck my palm but that its head has struck the nail. In another sense, of course, I am highly alert to the feelings in my palm and fingers holding the hammer. They guide my handling of it effectively, and the degree of attention that I give to the nail is given to these feelings to the same extent, but in a different way. The difference may be stated by saying that these feelings are not watched *in themselves* but that I watch something else by keeping aware of them. I know the feelings in the palm of my hand *by relying on them for attending to the hammer hitting the nail.* I may say that I have a *subsidiary* awareness of the feelings in my hand which is merged into my *focal awareness* of my driving the nail.

Let us think now of a probe instead of a hammer. A probe is used for exploring the interior of a hidden cavity. Think how a blind man feels his way by use of a stick, transposing the shocks transmitted to his hand and to the muscles holding the stick into an awareness of the things touched by the point of the stick. In the transition from hammer to probe we have the transition from practical to descriptive knowing, and we can see how similar the structures of the two are. In both cases we know something focally by relying subsidiarily on our awareness of something else.

This is also the way we know in perception. There are numberless items that contribute to my seeing my hand in front of me. They are known to me mainly by observing the various deficiencies caused by cutting out these several items. I can cut out marginal clues by looking at my hand through a blackened tube. I then observe that when I bring my hand closer to my eyes, while looking at it through the tube, my hand appears to become larger. The absence of marginal clues cut out by the tube prevents me from seeing my hand come nearer. Instead I see it *grow*. We can observe that the action of our eye muscles contributes clues to our vision if we apply drugs to these muscles that increase the effort needed to contract them. Since we regularly require more effort to see small objects, our subsidiary awareness of the increased effort becomes a clue to the perception of an object as small; and so the greater effort required for us to see, due to the effect of the drugs on our eye muscles, gets projected into the focal perception of an object as smaller than we otherwise know it to be. We are able also to notice that afflictions of the inner ear cause the whole spectacle before us to lose its stability and appear to roll. Finally, there is ample evidence that past experiences, which we can hardly recall, affect the way we see things.

Perception thus constitutes an observation of external facts without recourse to formal argument and even without any explicit statement of the result. We shall soon see more particularly how science, too, is based on such nonexplicit knowing, but first we must set out the general structure of such knowing and expose the indeterminacies introduced by it into empirical knowledge.

Consider the act of viewing a pair of stereoscopic pictures in the usual way, with one eye on each of the pictures. Their joint image might be regarded as a whole, composed of the two pictures as its parts. But we can get closer to understanding what is going on here if we note that, when looking through a stereo viewer, we see a stereo image at the focus of our attention and are also aware of the two stereo pictures in some peculiar nonfocal way. We seem to look through these two pictures, or past them, while we look straight at their joint image. We are indeed aware of them only as guides to the image on which we focus our attention. We can describe this relationship of the two pictures to the stereo image by saying that the two pictures function as *subsidiaries* to our seeing their *joint* image, which is their joint meaning. This is the typical structure of tacit knowing, which we shall now describe in some detail.

The grounds of all tacit knowing are items or particulars, like the stereo pictures, that we are aware of in the act of focusing our attention on something else, away from them. This is the *functional relation* of subsidiaries to the focal target, and we may also call it a *from-to relation*. Moreover, we can say that this relation establishes a *from-to knowledge* of the subsidiaries—a knowledge of them as they appear functionally in establishing the object of focal attention. Tacit knowing is a from-to knowing. Sometimes it can also be called a from-at knowing.

It will not be difficult to demonstrate the indeterminacies inherent in from-to knowledge, but let us first add some other features to the structure of from-to knowledge. A characteristic aspect of from-to knowledge is exemplified by the change of appearance which occurs when the viewing of a pair of stereo pictures transforms them into a stereo image. A stereo image has a marked depth and also shows firmly shaped ''solid'' objects not present as such in the original pair. It therefore involves us in a novel sensory experience, which has obviously been created by tacit knowing. Such *phenomenal trans-formation* is a characteristic feature of from-to knowing. In this manner the coherence we see in nature has an actually new *sensory* quality not possessed by the sense perceptions from which it is tacitly created.

A moment ago we anticipated another feature of from-to knowing when we said that the stereo image is the *joint meaning* of the stereo pictures. The subsidiaries of from-to knowing bear on a focal target, and whatever a thing bears on may be called its meaning. Thus the focal target on which they bear *is the meaning* of the subsidiaries. We may call this an act of sense-giving and recognize it as the *semantic aspect* of from-to knowing.

A few examples will help to make us familiar with these three aspects of tacit knowing: the functional, the phenomenal, and the semantic. We could discuss at this point both visual perception and scientific discovery, which are the main interest of this exposition, but they have complications which it is best to avoid for a time.

Take, as a simpler case, the from-to structure of the act of reading a printed sentence. The sight of the printed words guides our focal attention away from the type to a focal target that is its meaning. We have here both the *function* of from-to knowing and its *semantic* aspect. Its *phenomenal* aspect is also easy to recognize; it lies in the fact that a word in use *looks different* from the way it does to someone who meets it as a totally foreign word. The familiar use of a word, which is our subsidiary awareness of it, renders it in a way bodiless or, as is sometimes said, transparent.

Another example is the familiar case of tactile cognition: of using a probe to explore a cavity, or a stick to feel one's way in the dark. Such exploration is a from-to knowing, for we attend subsidiarily to the feeling of holding the probe in the hand, while the focus of our attention is fixed on the far end of the probe, where it touches an obstacle in its path. This perception is the function of tacit knowing and is accompanied by a particularly interesting *phenomenal transformation*. The sensation of the probe pressing on fingers and palm, and of the muscles guiding the probe, is lost, and instead we feel the point of the probe as it touches an object. And, in addition to the functional and the phenomenal, the probing has, of course, a *semantic* aspect, for the information we get by feeling with the point of the instrument is the meaning of our tactile experiences with the probe: it tells us what it is that we are observing by the use of the probe.

This use of probes may remind us of the fact that all sensation is assisted by some (however slight) *skillful* performance, the motions of which are performed with our attention focused on the intended action so that our awareness of the motions is subsidiary to the performance. From-to structure includes all skillful performances, from walking along a street to walking a tightrope, from tying a knot to playing a piano.

Let us confront these facts now with the further fact that there is one single thing in the world we normally know only by relying on our awareness of it for attending to other things. Our own body is this unique thing. We attend to external objects by being subsidiarily aware of things happening within our body. The localization of an object in space is based on a slight difference between the two images thrown on our retinas, the accommodation of our eyes, and on our control of our eye motion, supplemented by impulses received from the inner ear, which vary according to the position of the head in space. We become aware of all these things *only in terms of* our localization of the objects we are gazing at, and in this sense we may be said to be only subsidiarily aware of them. We may say, in fact, that to know something by relying on our awareness of it for attending to something else is to have the same kind of knowledge of it that we have of our body by living in it. It is a manner of being or existing.

Our subsidiary awareness of tools and probes can be regarded, then, as a condition in which they form part of our body. The way we use a hammer or a blind man uses a stick shows that in both cases we shift outward the points at which we make contact with things outside ourselves. While we rely on a tool or a probe, these instruments are not handled or scrutinized as external objects. Instead, we pour ourselves into them and assimilate them as part of ourselves.

We may generalize this to include the acceptance and use of the intellectual tools offered by an interpretative framework, in particular by the textbooks of science. While we rely on a scientific text, the text is not an object under scrutiny but a tool of observation. For the time being we have identified ourselves with it; and as long as our critical faculties are exercised in a from-to way by relying on this text, we shall continue to strengthen our uncritical acceptance of it.

There is no mystery about this. You cannot use your spectacles to scrutinize your spectacles. A theory is like a pair of spectacles; you examine things by it, and your knowledge of it lies in this very use of it. You dwell in it as you dwell in your own body and in the tools by which you amplify the powers of your body. It should be clear that someone who is said to be "testing" a theory is in fact relying, in this subsidiary and very uncritical way, upon other tacitly accepted theories and assumptions of which he cannot *in this action* be focally aware.

This conception of knowledge through indwelling will help to forge the final link between science and the humanities. Before we approach that point, however, we must further enlarge our scheme of personal knowledge to include both the kind of everyday knowledge we have of plants and animals, of life and death, of health and sickness, and the kind of expert knowledge which students of biology and medicine acquire in the laboratory and the clinic. We shall achieve this by observing that the two different kinds of awareness which we found interwoven in the use of a hammer or a probe are present in the same way in our awareness of any set of particulars perceived as a whole.

Consider any practical skill. It consists in the capacity for carrying out a great number of particular movements with a view to achieving a comprehensive result. The same applies to skillful knowledge, like that of a medical diagnostician; he too comprehends a large number of details in terms of a significant entity. In both kinds of skillful knowing we are aware of a multitude of parts in terms of a whole by dwelling in them. The two kinds of skillful knowing are actually always interwoven: a skillful handling of things must rely on our understanding them; and, on the other hand, intellectual comprehension can be achieved only by the skillful scrutiny of a situation. The kinship between the process of tool-using and that of perceiving a whole has in fact been so well established already by gestalt psychology that it may be taken for granted here without further argument.

Thus the structure of tacit knowing, as we have seen, includes a joint pair of constituents. Subsidiaries exist as such by bearing on the focus *to* which we are attending *from* them. In other words, the functional structure of from-to knowing includes jointly a subsidiary "from" and a focal "to" (or "at"). But this pair is not linked together of its own accord. The relation of a subsidiary to a focus is formed by the *act of a person* who integrates one to the other. The from-to relation lasts only so long as a person, the knower, sustains this integration.

This is not merely to say that if we no longer look at a thing we shall cease to see it. There is a specific *action* involved in dissolving the integration of tacit knowing. Let us describe this action.

We have seen that there are three centers of tacit knowledge: first, the subsidiary particulars; second, the focal target; and third, the knower who links the first to the second. We can place these three things in the three corners of a triangle. Or we can think of them as forming a triad, controlled by a person, the knower, who causes the subsidiaries to bear on the focus of his attention.

We can then say that the knower integrates the subsidiaries to a focal target, or we can say that for the knower the subsidiaries have a meaning which fills the center of his focal attention. Hence the knower can dissolve the triad by his own specific action: by merely looking differently at the subsidiaries. The triad will disappear if the knower shifts his focal attention away from the focus of the triad and fixes it on the subsidiaries.

For example, if, instead of looking at the stereo pictures through the viewer, we take them out and look at them directly, we lose sight of the joint appearance on which we had focused before. Or if we focus our attention on a spoken word and thus see it as a sequence of sounds, the word loses the meaning to which we had attended before. Or again, we can paralyze the performance of a skill by turning our attention away from its performance and concentrating instead on the several motions that compose the performance.

These facts are common knowledge, but their consequences for the theory of tacit knowledge are remarkable. For the facts confirm our view that we can be aware of certain things in a way that is quite different from focusing our attention on them. They prove the existence of two kinds of awareness that are mutually exclusive, a *from-awareness* and a *focal awareness*. They also confirm that in our from-awareness of a thing we see it as having a meaning, a meaning which is wiped out when we focus our attention *on* the thing of which we have only had a from-awareness. This thing will then face us in itself, in its raw bodily nature. The dual nature of awareness is made manifest here by the sense deprivation involved in substituting a focal awareness for a from-awareness.

It should be noted that this dual awareness is *not* due to the fact that we *cannot* become focally aware of all the subsidiary clues entering into an integrated meaning. Suppose that it would be possible, at least in principle, to identify all the subsidiaries involved in achieving a particular focal integration. We would still find that anything serving as a subsidiary ceases to do so when focal attention is directed on it. It turns into a different kind of thing, *deprived of the meaning* it had while serving as a subsidiary. Thus subsidiaries are—for this reason and *not* because we cannot find them all—*essentially* unspecifiable. We must distinguish, then, between two types of unspecifiability of subsidiaries. One type is due to the difficulty of tracing the subsidiaries—a condition that is widespread but not universal; the other type is due to a sense deprivation which is *logically* necessary and in principle absolute.

If this analysis convinces us of the presence of two very different kinds of awareness in tacit knowing, it should also prevent us from identifying them with conscious and unconscious awareness. Focal awareness is, of course, always fully conscious, but subsidiary awareness, or from-awareness, can exist at any level of consciousness, ranging from the subliminal to the fully conscious. What makes awareness subsidiary is its functional character, and we shall therefore claim the presence of subsidiary awareness even for functions inside our body, at levels inaccessible to us through direct experience.

As we have seen, subsidiaries function as such by being integrated to a focus on which they bear. This integration, because it is the tacit *act* of a person, can be either valid or mistaken. Perception can be true or mistaken—that is, we may later be able to judge it as one or the other—and so also can our judgments of coherence, whether in deciding the facts of a legal case or when perceiving coherence in nature. To arrive at such conclusions may therefore legitimately be called *acts of tacit inference*. Such inferences differ sharply, however, from drawing conclusions by an explicit deduction.

Piaget has described the contrasts between a sensorimotor act like perception and a process of explicit inference. Explicit inference is reversible: we can go back to its premises and go forward again to its conclusions, and we can rehearse this process as often as we like. This is not true for perception. For example, once we have seen through a visual puzzle, we cannot return to an ignorance of its solution. This holds, with some variations, for all acts of tacit knowing. We can go back to the two pictures of a stereo image by taking them out of the viewer and looking at them directly, but this completely destroys the stereo image. When flying by airplane first started, the traces of ancient sites were revealed in fields over which generations of country folk had walked without noticing them. But, once he had landed, the airplane pilot could no longer see them either.

In some cases sense deprivation is not permanent. Thus we can concentrate on the sound and the action of our lips and tongue in producing a word, and this will cause us to lose the meaning of the word, although the loss can be instantly made good by casting the mind forward to the saying of something that makes use of the word. The same is true for a pianist who paralyzes his performance by intensely watching his own fingers; he too can promptly recover their skillful use by attending once more to his music. In these instances the path to the integrated relation—which may originally have taken months of labor to establish—is restored from its abeyance in a trice; in the same moment, the sight of the subsidiary particulars is lost.

Explicit inference is very different; no such breaking-up and rediscovery take place when we recapitulate the deduction of the theorem of Pythagoras. Once this basic distinction between explicit inference and tacit integration is clear, it throws new light on a 100-year-old controversy. In 1867 Helmholtz offered to interpret perception as a process of unconscious inference, but this theory was generally rejected by psychologists, who pointed out that optical illusions are not destroyed by demonstrating their falsity. Psychologists had assumed, quite reasonably, that "unconscious inference" had the same structure as a conscious explicit inference. But if we identify "unconscious inference" with tacit integration, we have a kind of inference that is not damaged by adverse evidence, as explicit inference is. This difference between a deduction and an integration lies in the fact that deduction connects two focal items, the premises and consequents, while integration makes subsidiaries bear on a focus. Admittedly there is a purposive movement in a deduction—which *is its essential tacit coefficient*; but the deductive *operation* can be mechanically performed, while a tacit integration is intentional throughout and, as such, can be carried out *only by a conscious act of the mind*. Brentano has taught that consciousness necessarily attends to an object and that only a conscious mental act *can* attend to an object. Our analysis of tacit knowing has amplified this view of consciousness. It tells us not only that consciousness is intentional but also that it always has roots *from which* it attends to its object. It includes a tacit awareness of its subsidiaries.

Such integration cannot be replaced by any explicit mechanical procedure. In the first place, even though one can paraphrase the cognitive content of an integration, the sensory quality which conveys this content cannot be made explicit. It can only be lived, can only be dwelt in.

The irreducibility of tacit integration can be observed more fully in practical knowing. We can point to our incapacity to control directly the several motions contributing to a skillful performance, even in such a familiar act as using our limbs. But these cases are perhaps too common to stir the imagination. So let us take instead the example of finding one's way about while wearing inverting spectacles. It is virtually impossible to get about while wearing inverting spectacles by following the instruction that what—in the case of right-left inversion—is seen on the right is actually on the left, while, conversely, things seen on the left are on the right. The Austrian psychologist, Heinrich Kottenhoff, however, has shown that continued effort to move about while wearing inverting spectacles produces a novel quality of feeling and action: by integrating the *inverted* images to appropriate sensorimotor responses, the subject again finds his way about. This reintegration can be performed only subsidiarily; any explicit instruction to reintegrate images and sensorimotor responses would be quite meaningless.

For the same reasons we cannot learn to keep our balance on a bicycle by trying to follow the explicit rule that, to compensate for an imbalance, we must force our bicycle into a curve—away from the direction of the imbalance—whose radius is proportional to the square of the bicycle's velocity over the angle of imbalance. Such knowledge is totally ineffectual unless it is known tacitly, that is, unless it is known subsidiarily—unless it is simply dwelt in.

The way in which particulars are picked up and assimilated as subsidiaries will be dealt with when we speak of scientific discovery. But we may anticipate one of its aspects, pointed out by Konrad

Lorenz, who arrived independently at the view that science is based on a gestalt-like integration of particulars. He demonstrated that the speed and complexity of tacit integration far exceeds the operations of any explicit selection of supporting evidence. I have recently developed this idea further by showing that the serial behavior demonstrated by Lashley in the production of a spoken sentence can be understood by considering that tacit knowing can pick up simultaneously a whole set of data and combine them in a meaningful spoken sequence.

An integration established in this summary manner will often override single items of contrary evidence. It can be damaged by new contradictory facts *only if these items are absorbed in an alternative integration* which disrupts the one previously established. In Ames's skewed-room experiment, the illusion that a boy is taller than a man persists as long as the evidence of the room's skewed shape is slight; but the illusion is instantly destroyed if, through shifting his positon or by tapping the ceiling, the observer becomes more aware of the room's skewedness. This new awareness effectively competes with the hitherto established integration and destroys it.

David Hume maintained that, when in doubt, we must suspend judgment. This theory might apply to conclusions derived on paper, conclusions that can be manipulated at will—at least on paper. But our eyes continue to *see* a little boy taller than a grown man (although we know this must be false) as long as no feasible alternative integration is presented to our imagination. When we are presented with an alternative that appears to us to be more meaningful and so more true, a new perception will take place and correct our errors. Helmholtz could have answered his opponents on this score, but he would have had to admit that his "unconscious inferences" were very different from conscious inferences.

The characteristic structure of all our personal knowledge comes out even more vividly when we realize that all knowing is action—that it is our urge to understand and control our experience which causes us to rely on some parts of it subsidiarily in order to attend to our main objective focally. As we watch the operation of this urge, we shall see emerging another important feature of personal knowledge. It will appear that all personal knowing is intrinsically guided by impersonal standards of valuation set by a self for itself.

Take first the process of mastering a skill. Here the emphasis of our knowing lies on producing a result. The effort involved in acquiring knowledge and skillfully applying it may then be said to be guided by a purpose. It is in the light of this purpose that certain things are made to serve us as tools and that certain movements of our body are skillfully coordinated. The economical and effective achievement of this purpose sets a standard for our skill. It is by striving to reach this standard that we pick up in practice, usually without any focal awareness of doing so, the elements of a successful performance. Thus our efforts to achieve a skill make it possible for us to say that we are being guided by a purpose.

When, on the other hand, the emphasis of our knowing lies in recognizing or understanding a thing, the effort involved in acquiring such knowledge may be said to be guided by our attention. A biologist, a doctor, an art dealer, and a cloth merchant acquire their expert knowledge in part from textbooks, but these texts are of no use to them without the accompanying training of the eye, the ear, and the sense of touch. Only by attentively straining their senses can they acquire the right sense, or feel, for identifying a certain biological specimen, the symptoms of a certain sickness, a genuine painting by a certain master, or a fabric of a distinctive quality. By such training the expert develops an exceptional fastidiousness which enables him to act as an appraiser of the value or meaning of certain objects or conditions.

So every act of personal knowing contributes toward establishing an appropriate standard of excellence. Athletes or dancers putting forward their best are acting as critical experts of their own performances; but whether applying standards to themselves or not, experts are the acknowledged critics of certain things. And when a person is acknowledged as an expert, he is believed to know whether such things fulfill the standard of good specimens of their own kind.

Thus the observer's participation in the act of knowing leads to a point where observation assumes the functions of an appraisal by standards which the observer regards as impersonal, i.e., generally applicable rather than personally idiosyncratic. Even courts of law will rely on the expert's capacity to appreciate the true presence of a degree of ingenuity in a new device, and they will base the granting of a patent on this appreciation. Mathematics can also be said to exist as a science only if we trust our capacity to appreciate the actually existing profundity and ingenuity of certain processes of inference.

We are now ready for the final step, which combines personal indwelling with an appraisal according to standards accepted as universal. It is clear that a naturalist appreciates a healthy plant or animal by a standard to which he attributes universality. They are "right" *after their kind*. He appreciates the coherent behavior of animals in the same way. His appreciation is based here on entering into another individual's purpose and action. Feats of intelligence can be observed only if we dwell in their parts *as being* intelligently integrated, thus identifying ourselves (in this sense) with the person whose intelligence we appraise. Our capacity for making sense of, for understanding, another person's action by entering into his situation and by judging his actions from within his own point of view thus appears to be but an instance of the technique of personal knowing, the elements of which we earlier identified by the way we dwell in subsidiary elements in order to read a map or drive a nail with a hammer or grope our way in the dark by using a stick.

This conclusion fulfills at least part of our program. We have seen that the shortcomings of the Laplacean ideal of science must be remedied by acknowledging our personal knowing—our indwelling—as as integral part of all knowledge. This amendment, we can now see, bridges the gap between the natural sciences and the study of man. Having recognized personal participation as the universal principle of knowing and having determined the structure of this knowing, we are now able to see that the personal participation through which we reach our evaluation of human actions as the actions of sentient, intelligent, and morally responsible beings is a legitimate instance of scientific knowing. It is the way in which we dwell in these particular subjects in achieving our knowledge of them—our understanding of their actions. It is thus not at all "unscientific." *All* knowing is personal knowing—participation through indwelling.

We may now go further. We have seen that our personal knowing operates by an expansion of our person into a subsidiary awareness of particulars, an awareness merged with our attention to a whole, and that this manner of living in the parts results in our critical appraisal of their coherence. We may therefore also accredit our living within a historical situation and our acceptance of a certain role in it as legitimate guides to our responsible participation in the problems presented by these situations. Science will then no longer seem to us to require that we study man and society in a *detached* manner, and we shall have been restored to an acceptance of our position as human members of a human society.

Something more vital, therefore, follows from formulating tacit knowing as an act of indwelling, as *personal* knowledge. It deepens our knowledge of *living* things. Biology studies the shapes of living things and the way they grow into these shapes from germ cells; it describes the organs of plants and animals and explains the way they function; it explores the motor and sensory capacities and the intelligent performances of animals. These aspects of life are all controlled by biological principles. Morphology, embryology, physiology, psychology, all study the principles by which living beings form and sustain their coherence and respond in continually new ways to an immense variety of novel circumstances. The ways in which animals, such as cattle or dogs or men, coherently control their bodies all form comprehensively functioning systems. Any chemical or physical study of living things that is irrelevant to these workings of the organism forms no part of biology.

We therefore recognize and study the coherence of living things by integrating their motions—and any other normal changes occurring in their parts—into our comprehension of their functions. We integrate mentally what living beings integrate practically—just as chess players rehearse a master's game to discover what he had in mind. We share the purpose of a mind by dwelling in its actions. And so, generally, we also share the purposes or functions of any living matter by dwelling in its motions in our efforts to understand their meaning.

3

RECONSTRUCTION

Now at last we are in a position to show how philosophic thought and the methodological principles of science have been misguided by not having achieved a clear knowledge of tacit knowing. Let us take as examples, first, the knowledge of other minds; second, the mind-body problem; third, universal terms; fourth, principles of explanation; fifth, empirical generalization; and sixth, scientific discovery.

Knowledge of Other Minds

Up to a point, Gilbert Ryle has argued about our knowledge of other minds as we have just done in the preceding chapter. He has strikingly demonstrated that we do not get to know the workings of another mind by a process of inference. We are rather following them. Ryle, however, had no conception of tacit knowing, so he concluded that body and mind are "not two different species of existence," "not tandem operations." Accordingly, he held that "intelligent performances are not clues to the workings of minds; they are those workings."[1]

This theory we must reject. The theory of tacit knowing, while it also tells us that we do not know another mind by a process of inference, nevertheless retains the dualism of mind and body in this sense: it says that the body seen focally is one thing, while the body seen subsidiarily points to another thing; these two different things are

Part of this chapter is adapted from Michael Polanyi, "Logic and Psychology," *The American Psychologist* 23 (January 1968): 34–42. Copyright © 1968 by the American Psychological Association. Reprinted by permission.

the body and the mind. We shall discuss this mind-body question more fully in the next section of this chapter.

Merleau-Ponty, writing some years before Ryle, anticipated his premises by declaring: "I do not understand the gestures of others by some act of intellectual interpretation . . ."; to which he added, a little further on: "The act by which I lend myself to the spectacle must be recognized as irreducible to anything else."[2]

Merleau-Ponty anticipated the existential commitment present in tacit knowledge but did so without recognizing the triadic structure which determines the functions of this commitment—the way it establishes our knowledge of a valid coherence. The contrast between explicit inference and an existential experience imbued with intentionality is not sufficient for defining the structure and workings of tacit knowing. We are offered an abundance of brilliant flashes without a constructive system.

The philosophy of behaviorism does not contest the duality of body and mind,[3] but it assumes that all mental performances can be fully specified without referring to any mental motives. Can this be true? Consider an analogy. All textbooks of physiology refer to organs and to the fuctions of organs, and in spite of frequent solemn declarations that such teleological conceptions are unnecessary and indeed objectionable, no one has yet published a textbook of physiology that does not speak of organs and their functions. For the biological functions of organs can be known only as parts of their meaningful combination. To describe, as behaviorists claim to do, the workings of the mind without relying on the guidance of mental motives is as impossible as it is to describe physiological events occurring in an organ without being guided by the observation of its coherent physiological function.

To some extent behaviorism succeeds in claiming the impossible, because our knowledge of the mind's coherence is so stable that any isolated workings of the mind will instantly evoke the underlying mental motives which the isolated workings are supposed to replace. In *Personal Knowledge* I have described how the behaviorist psychology of learning is able to achieve plausibility.[4] It does so first by limiting the inquiry to the crudest forms of learning. These much simplified cases can then be plausibly described in objectivist terms. Even so, they can be shown to apply to the actual process of learning only because their meaning is tacitly understood by their bearing on the mental events covertly kept in mind. One thus succeeds in using objectivist terms (which are, strictly speaking, meaningless) as effective paraphrases for the mentalistic terms they are supposed to eliminate.

Noam Chomsky's critique of B. F. Skinner's *Verbal Behavior* presents many illustrations of such behaviorist paraphrasing.[5] Apparently objective terms, like "stimulus," "control," "response," and so on, are so used that their ambiguity covers the mental terms they are supposed to replace. Chomsky shows that either you use these objectivist terms literally—and then what is said is obviously false and absurd—or you use them as substitutes for the terms they are supposed to eliminate—and then you do not say anything else than what you would have said in mental terms.

The same criticism applies to the mechanical simulation of tacit knowing. A tacit integration can be only superfically paraphrased by a computer, and such paraphrases succeed by evoking the mental qualities which they claim to enact.

The fact is, we know other minds by *dwelling in* their acts—as the chess player comes to know the mind of the master whom he is studying. He does not *reduce* the master's mind to the moves the master makes. He dwells in these moves as subsidiary clues to the strategy in the master's mind which they enable him to see. The moves become meaningful at last only when they are seen to be integrated into a whole strategy. And a person's behavior, in general, becomes meaningful only when integrated into a whole mind.

The Mind-Body Problem

Turning now to the old mind-body problem, we see how our theory of tacit knowing can enable us to formulate the mind-body dualism as the disparity between the experience of a subject observing an external object like a cat and the experience of a neurophysiologist observing the bodily mechanisms by which the subject sees the cat. The two experiences are very different. The subject sees the cat but does not see the mechanism he uses in seeing the cat, while the neurophysiologist sees the mechanism used by the subject but does not share the subject's sight of the cat.

Admittedly, the neurophysiologist is aware of the subject's mind seeing the cat, for up to a point he shares the subject's mind by dwelling in its external workings. He sees the subject as a sentient, thinking fellow man and can understand his response to the sight of the cat; for example, he can follow the subject's description of the cat.

And, of course, on his part, the subject can understand the description of his own neural mechanism, although he himself cannot observe it.

But the fact remains that to see a cat differs sharply from a knowledge of the mechanism of seeing a cat. They are each a knowledge of quite different things. The perception of an external thing is a from-to knowledge. It is a subsidiary awareness of bodily responses evoked by external stimuli, seen with a bearing on their meaning situation at the focus of our attention. The neurophysiologist who is focusing his attention on these bodily responses does not experience their integration. He has an at-knowledge of the subject's body, with its bodily responses at the focus of his attention. These two experiences have a sharply different content, and the difference represents the viable core of the traditional mind-body dualism. "Dualism" thus becomes merely an instance of the change of subject matter which occurs when one shifts one's attention from the direction on which the subsidiaries bear and focuses instead on the subsidiaries themselves.

Current neurophysiology is based on a parallelism of body and mind as two aspects of the same thing. This theory is false; but it is plausible, because body and mind are obviously closely connected. The mind relies for its actions on the body; and since, to our modern thinking, matter appears more substantial than mind, it seems reasonable to modern thinking to assume that the body, being the more substantial of the two, altogether determines mental actions.

We must contradict this, however briefly, by defining the actual relation between body and mind. Some principles—for example, those of physics—apply in a variety of circumstances. These circumstances are not determined by these principles; they are the boundary conditions for the operation of these principles, and no principles can determine their own boundary conditions. When there is a principle controlling the boundary conditions of another principle, the two must operate jointly. In this relation, the first (the one controlling the boundary conditions) can be called the higher principle; the second, the lower principle.

Such a two-level structure can be seen to operate in all sorts of things. In the playing of chess, for instance, the conduct of the game is an entity directed by a stratagem—a stratagem which, of course, must rely upon the observance of the particular rules of chess. This relation does not hold in reverse, for the rules of chess do not *direct* any particular strategy. They leave open an infinite range of stratagems. Moves in chess are therefore meaningless by themselves. Their meaning lies in serving jointly the performance of a stratagem, and this in turn relies for its meaning upon observance of the rules of the game.

All these relations become clearer in the case of a skill which comprises a number of levels in the form of a hierarchy. The production of an oral communication, for instance, includes five levels. At the lowest level (1) there is the production of voice sounds. These sounds are combined at the next-higher level (2) into the utterance of meaningful words. These words then (3) achieve further meaning by our dwelling in them in order to integrate them into the sort of meaning that only sentences can have. Sentences themselves are then (4) worked together into a style or mode of creating impressions or intelligible points—something that does not exist in sentences as such. Finally (5), the style or mode of creating impressions must itself be used (dwelt in) toward the attainment, *through* it, of the ideas or results that are the ultimate focal aim of the communication. Without the last level the communication would not really *say* anything, even though the other levels remained fully existent. What we would be left with would be an incoherent rhapsody of impressions or points, full of sound and fury (perhaps), but signifying nothing. We would have to say: Well, yes, but what does it all *mean*?

It is clear from this example that the principles of each level operate under the control of the next-higher level. The voice you produce is shaped into words by a vocabulary; a given vocabulary is shaped into sentences in accordance with a grammar; and the sentences are fitted into a style, which in its turn is shaped by our efforts to convey the ideas of the composition. Thus each level is subject to dual control: first, by the laws that apply to its elements in themselves and, second, by the laws that control the comprehensive entity formed by them.

Such multiple control is made possible again by the fact that the principles governing the isolated particulars of a lower level leave indeterminate their boundary conditions. These will be controlled by a higher principle. Voice production leaves largely open the combination of sounds into words, which is controlled by a vocabulary. Next, a vocabulary leaves largely open the combination of words to form sentences, which is controlled by a grammar; and so the sequence goes on. Consequently, the operations of a higher level cannot be *accounted for* by the laws governing its particulars, which form the next-lower level. You cannot derive a vocabulary from phonetics; you cannot derive a grammar from a vocabulary; a correct use of grammar does not account for good style; and a good style does not supply the *content* of an oral communication.

Mental principles and the principles of physiology form a pair of such jointly operating principles. The mind relies for its workings on the continued operation of physiological principles, but it controls the boundary conditions left undetermined by physiology. This lends substance to the conclusion we derived from the structure of tacit knowing, that body and mind are profoundly different; they are not two aspects of the same thing.[6]

This should dispose of any grounds for a theory of mechanical determinism of the mind by the body. If mind and body were two aspects of the same thing, the mind conceivably could not do anything but what the bodily mechanism determined. But the existence of two kinds of awareness—the focal and the subsidiary—distinguishes sharply between the mind as a lived-in, from-to experience, and the subsidiaries of this experience when these are viewed focally as a bodily mechanism. In addition, the necessity that the boundary conditions limiting the operation of any set of lower-level principles must be different from these principles means that the mind can readily be understood to serve as a set of such boundary conditions for the operation of the laws of neurophysiology. Though rooted in the body, the mind is therefore free in its actions from bodily determination—exactly as our common sense knows it to be free.

Universal Terms

The mere fact that biology deals with classes of plants and animals involves indeterminacies which, for centuries past and up to this day, philosophers have tried in vain to eliminate.

Plato and his school were the first to be troubled by the fact that *in applying our conception of a class of things we keep identifying objects that are different from one another in every particular*. If every man is clearly distinguishable from every other man and we yet recognize each of them as a man, what kind of "man" is the one by reference to whom all these men are recognized? He cannot be both fair and dark, young and old, or brown, white, black, and yellow all at once; but neither can he have any one of these alternative properties or, indeed, any particular property whatever. Plato concluded that the general idea of man refers to a *perfect man*, who has no particular properties and of whom individual men are imperfect copies, corrupted by having such properties.

The Platonic idea of man, however, embodies instead of eliminates the paradox of identifying different individuals. An attempt to avoid this difficulty was made about 900 years ago by Roscellinus through his proposal of nominalism. According to this doctrine, the word "man" is but the name of a collection of individual men. But the indeterminacy reappears once more when we ask how to justify the labeling of a collection of different individuals by the same name and how, moreover, we can continue to label in a constant fashion, as time goes on, any number of further individuals differing in every particular from any individual thus labeled before and yet can continually exclude a vast number of other individuals as not belonging to the class we have labeled.

Perhaps it is no wonder that, another 700 years later, Kant should have declared that the way our intelligence forms and applies the conception of a class of particulars "is an art concealed in the depths of the human soul whose real modes of activity nature is hardly likely ever to allow us to discover."[7]

Yet, in the present age of philosophy, in 1945, F. Waismann attempted to answer the question deemed insoluble by Kant. He pointed out that general terms have an "open texture" which admits differences in the instances to which it applies.[8] But this merely shifts the question, for to ascribe "open texture" to a word is merely to imply that among an indefinitely extending series of different objects the word properly applies to some objects and not to the rest. The question how this is done remains open, exactly as before.

No wonder that the more incisive reappraisal of the old quandary by W. V. O. Quine led him back to the conclusion reached by Kant: that the grounds on which instances are assigned to an empirical category of objects are inscrutable.[9]

Yet the speculations of two and a half millennia, striving to eliminate the indeterminacy involved in subsuming a presumed instance of a class under that class, seem to have been misguided. This indeterminacy is irreducible, but its comprehension is safely controlled by tacit integration.

Tacit knowing commonly integrates groups of particulars into their joint meaning. Members of a class such as a species, a family, or a language—or members of any other group properly denoted by a single universal term—possess a joint focal meaning when dwelt in as subsidiary clues to such a meaning, even though their focus is almost empty in contrast to a focal object of perception. Moreover, the meaning of a class is an aspect of reality, for it points to yet unrevealed joint properties of its members. If the joint appearance of disparate members of a class in the conception of a class should need support by analogy, think once more of binocular vision uniting two slightly different images in a single image of a different sensory character; or of the fact that the way we see an object integrates, among many other events in the body, innumerable memories beyond conscious recollection; or of a metaphor fusing two disparate ideas in a powerful joint meaning we have never before encountered.

Principles of Explanation

Once we have fully realized the structure of the tacit triad, the recognition and acceptance of its functions become irresistible. We shall demonstrate this by discussing how the way in which we produce a scientific explanation depends on our propensity to be puzzled. We must note, to begin with, that being puzzled implies a selective judgment. For example, most physicists believe today that it is nonsensical to be puzzled by the fact that one radium atom decomposes today and another perhaps fifty years later, since there can be no explanation for this. Some scientists, however, oppose this view, and this opposition implies a fundamental difference of opinion. Furthermore, it is thought to be useless to be puzzled by events which, although they may be explicable in principle, are not ripe for explanation or are not worth the trouble of explaining them. There are many variations of the advisability of being puzzled. But, in any case, a scientific explanation must serve to dispel puzzlement.

Relief from puzzlement, however, may be attained by other means than that of explanation, and such cases throw light on explanation itself. Suppose we are puzzled by the way an intricate piece of machinery is constructed and the way it works or by the layout of a building in which we keep losing our way. What we are seeking here is an understanding of the machine or the building—an insight into them, but not an explanation. Such insight is a particular type of tacit integration that has not yet been mentioned. Its subsidiary items are the particulars of the complex entity—the machine, or the rooms in the building; and when we integrate these particulars and thus bring out their joint meaning, their puzzling aspect is transformed into a lucid image. Our puzzlement in these cases is relieved by an insight which is itself simply our own meaningful integration of the parts of the complex entity.

Such insight is in general not an image that can be laid out on a sheet of paper, for it often serves to illuminate a three-dimensional spatial arrangement of opaque objects, like complex machines and buildings. The anatomy of a complex living thing, or a complex arrangement of geological strata, or a complex atomic arrangement in crystals may also be baffling and can be understood only by an act of insight. But any plane aspect can give only a fragmentary view of such systems; only by combining such aspects in the imagination can a three-dimensional understanding of the aggregate be achieved. Such insight is a purely mental fact, like any other focal target, comprising a large number of subsidiaries. Such insight differs, however, from all the focal targets (like those of stereo vision, reading a sentence, probing a cavity, etc.) that we have mentioned before, in that this focal target does not lie away from the subsidiaries but coincides in our imagination with its parts. It is the aggregate of its parts mentally seen as they are (or as we believe them to be). It signifies that the imaginative probing of a puzzling aggregate has established in it an intelligible coherence or meaning.

If this coherence can be formulated in explicit terms, it can amount to the kind of discovery exemplified in the discovery of Kepler's planetary laws. But its scope can be extended also in the opposite direction, down to the level of animal intelligence, as in the classic experiments of Wolfgang Köhler on the powers of insight of chimpanzees. The chimpanzee, presented with a string wound twice, loosely, around a rod, realized that the string could not be disengaged by pulling at its end but must first be unwound. Returning to human thought, we note that biology frequently starts with the question "How does it work?"—which is much like the way we start when we try to understand a piece of complex machinery. Furthermore, we have shown previously that we understand living things, the functioning of their organs and the working of their intelligence, by an act of indwelling, which is also an act of insight. Similar insight is involved in taxonomy, which orders biological specimens much as X-ray crystallography does in establishing the space group to which a crystalline specimen belongs.

The establishment of insightful coherence has many forms, some only tacitly known, others explicitly formulable, some in common use, and others forming part of science.

This brings us back to the subject of scientific explanation. The possibility for extending the recognition of inference in nature consists in subsuming a natural law within a more general law of which it is a special case. This procedure has been singled out by modern philosophers from J. S. Mill to Carl G. Hempel, Hempel and Oppenheim, and Ernest Nagel as constituting the scientific explanation of natural phenomena.[10] But thus to limit the study of a wide area to the analysis of a fragment of it obscures the subject. Michael Scriven, criticizing this analysis of explanation, has suggested that concepts condemned by many logicians ''as psychological not logical''—for example, understanding, belief, judgment—may have to be returned to circulation.[11] Our present account of relief from puzzlement by the spreading of coherence extends Scriven's criticism. To define the explanation of an event as its subsumption under a general law leaves unexplained its capacity to relieve puzzlement and isolates it from numerous other, more fundamental, acts which have this capacity. Explanation must thus be understood as a particular form of insight— an insight that relieves our puzzlement through the establishment of a more meaningful integration of parts of our experience, achieved through the subsumption of a natural law under a more general law.

The consequences of striving for a strictly formal account of ''explanation''—i.e., reducing ''explanation'' to any merely *formal* subsumption of a natural law under a more general law—are similar to the ones we saw in the case of behavioristic explanation. The actual subject matter is restricted to a fragment found suitable for formalization. This formalization, if carried out strictly, produces a result that is strictly, in itself, empty of any bearing on the subject matter; but by calling it an ''explanation,'' one imbues it with the memory of that informal, insightful act of the mind which it was supposed to replace. Such a denial of mental powers avoids its ultimate consequences by borrowing the qualities of the very powers it sets out to eliminate. This we may call pseudosubstitution. A pseudosubstitution is a gesture of intellectual self-destruction that is kept within safe bounds by its inconsistency. We have mentioned this practice in our criticisms of behaviorism and of the simulation of tacit integration by computers; our culture is pervaded by such intellectual subterfuges.

Empirical Generalization

If the orthodox theory of scientific explanation is misleading, the treatment of empirical generalization is equally so. Without going into detail, let us point out here three major errors which have resulted from the attempt to define empirical validity by strict criteria. First, since no formal procedure could be found for producing a good idea from which to start an inquiry, philosophers virtually abandoned the attempt to understand how this is done. Second, having arrived at the conclusion that no formal rule of inference can establish a valid empirical generalization, they denied that any such generalization can be derived from experimental data—while ignoring the fact that valid generalizations are commonly arrived at by empirical inquiries based on *informal* procedures. Third, they claimed that a hypothesis is formally refutable by a single piece of conflicting evidence—an illusory claim, since one cannot formally identify a contradictory piece of evidence.

I pointed out these mistakes in 1946, in *Science, Faith and Society*, and also developed there some ideas about the informal powers which guide the pursuit of science and provide criteria for accepting its results. Underlying these ideas was the assumption that science is based on our powers to discern coherence in nature. This discernment is what sight and other senses do on the physiological level in the act of perception, and I generalized these powers to include scientific discovery. I said that "the capacity of scientists to guess the presence of shapes as tokens of reality differs from the capacity of our ordinary perception, only by the fact that it can integrate shapes presented to it in terms which the perception of ordinary people cannot readily handle."[12]

Since that was written, I have tried to pursue systematically the kinship between perception and scientific discovery. Among earlier writers upon whom I relied for my work were Poincaré, Hadamard, and Polya.[13] Confirmation of my position came later in the works of philosophers criticizing the hypothetico-deductive method. Fuller confirmation was to be found in the admirable *Harvard Case Histories in Experimental Science* directed by J. B. Conant. Thomas S. Kuhn's book *On the Structure of Scientific Revolutions* brought further confirmation of my views in detail, and the analysis of science by Leonard K. Nash, *The Nature of the Natural Sciences*, combined my own views with those of other authors in a presentation of science as an insight into the nature of reality. A few years ago I experienced the wonderful surprise of finding my basic assumptions anticipated by the nineteenth-century philosopher, William Whewell.[14]

Scientific Discovery

In discussing the first five problems dealt with in this chapter we have been demonstrating the powers of tacit knowing with respect to certain long-standing epistemological difficulties relating more or less to what might be called the "statics" of knowledge, and so also of science. In turning now to a demonstration of how these powers can yield an explanation of the sixth problem as well, that of scientific discovery, we will be showing that the notion of tacit knowing can also bring the *dynamics* of science into its epistemological framework with ease and clarity. We shall see that tacit knowing is able to make good sense of an aspect of science that has flatly resisted all efforts to bring it into the ambience of strict formalization. As we deal with science in the making, from the sighting of a problem to the claiming of discovery, in terms of tacit knowing, we shall find further dynamic dimensions to tacit knowing that will also prove to be of great value to us in bringing science and the humanities together.

We have previously described how we can wipe out the meaning of a word by attending directly to its physical aspect. Such loss of meaning can be restored by casting the mind forward to the act of saying something in which the word will have to be included with its proper meaning.

This illustrates a basic principle of tacit dynamics. It also shows us that no instance of tacit knowing is truly static. The casting-forward of an intention is an act of the imagination. It is only the imagination that can direct our attention to a target that is as yet unsupported by subsidiaries. Although the lost meaning of a word is in recent memory when our imagination sallies forth, seeking to restore it, this meaning is not yet present. But the imagination must feel that this lost meaning is available if it is to start on an action that will require the use of the word in its restored meaning, for only by thrusting in a feasible direction can the imagination succeed in evoking a lost meaning.

Let us now develop further this kind of imaginative action. Our intention to say something normally evokes our verbal expression of it. Modern linguistics, however, has shown that most of our speech consists of sentences we have never before used; they are composed for the first time on this occasion. Yet when we start uttering such a sentence, we are as a rule confident that we shall find the words we need and bring them out in the order expressing our thought. We can speak as we do because we feel that many thousands of words are available for our novel purposes, and we can trust the powers of our imagination, bent on this purpose, *to evoke from these available resources the implementation of our purpose*, just as our intention to raise our arm evokes the coordination of our nerves and muscles in the accomplishment of this intention.

To listen to speech and understand novel sentences is a similar action. Our imagination moves ahead of a novel text before us, trusting itself to evoke an understanding of it, and it commonly does so by evoking and combining available meanings that will match the meaning of the unprecedented sentences before us.

This is, in essence, also the way the imagination works on its major tasks. It is the way it works in search of discovery. The scientist's imagination does not roam about, casting up random hypotheses to be tested. He starts by thrusting forward ideas he feels to be promising, because he senses the availability of resources that will support them; his imagination then goes on to hammer away in directions felt to be plausible, uncovering material that has a reasonable chance of confirming these guesses.

The scientific imagination, achieving discovery, passes through a typical life-cycle. The claim of Copernicus that his system was real declared a vision of boundless unknown implications. For sixty years this vision lay dormant in men's minds. Then Kepler's imagination transformed the vision of Copernicus into a new and dynamic anticipation. He brushed aside as unreal Copernicus's fantastic system of epicycles and set out to find laws reflecting the mechanical foundations of the heliocentric order. His triumph ushered in a new period of becalmed vision to the imagination. Thus ended, as it were, a life-cycle.

When the imagination goes into action to start a scientific inquiry, it becomes not only more intense but also more concrete, more specific. Although it thrusts toward a target that is as yet empty, the target is seen to lie in a definite direction. The target represents a particular problem, pointing vaguely, but still always pointing, to a hidden feature of reality. Once the problem is selected, its pursuit will rely on a particular range of resources that are felt to be available. These resources will include the amount of labor and money needed for the quest. No problem may be undertaken unless we feel that its possible solution would be worth the probable expense.

These strangely perceptive anticipations are not arrived at either by following rules or by relying on chance, nor are they guaranteed by a wealth of learning. Yet their use is common and indeed indispensable; no scientist can survive in his profession unless he can make such anticipations with a reasonable degree of success. To do this, he needs exceptional gifts. But his gifts, although exceptional, are the same kind that underlie the mere use of speech, and they are actually found at work in every deliberate human action. They are intrinsic to the dynamics of all from-to knowledge, down to the simplest acts of tacit knowing.

In science the path from problem to discovery can be lengthy. The inquiry having been launched, the imagination will continue to thrust forward, guided by a sense of potential resources. It batters its path by mobilizing these resources, occasionally consolidating some of them in specific surmises. These surmises will then tentatively fill, up to a point, the hitherto empty frame of the problem.

It is a mistake to think of heuristic surmises as well-defined hypothetical statements which the scientist proceeds to test in a neutral and indeed critical spirit. Hunches often consist essentially in narrowing down the originally wider program of the inquiry. They may be very exciting and may indeed turn out later to have been crucial; yet they are for the most part far more indeterminate than the final discovery will be. The range of their indeterminacy lies at some point between the indeterminacy of the original problem and the indeterminacy of its eventual solution.

Besides, the relation of the scientist to his surmises is one of passionate personal commitment. The effort that led to a surmise committed every fiber of his being to the quest; his surmises embody all his hopes.

Current theories of scientific inquiry that ignore the mechanism of tacit knowing must ignore and indeed deny such commitments. The tentativeness of the scientist's every step is then taken to show that he is uncommitted. But every step taken in the pursuit of science is definitive, definitive in the vital sense that it definitely disposes of the time, the effort, and the material resources used in taking that step. Such investments add up, with frightening speed, to the whole professional life of the scientist. To think of scientific workers cheerfully trying this and trying that, calmly changing course at each failure, is a caricature of a pursuit that consumes a man's whole person. Any questing surmise necessarily seeks its own confirmation.

We might suppose, then, that a problem is eventually resolved by the discovery of a coherence in nature whose hidden existence had first been sighted in a problem and had then become increasingly manifest during the pursuit of that problem. But this would leave out of account another factor, the identification of which we owe to Henri Poincaré. In a classic essay included in *Science et méthode* he described two stages in the way we hit upon an idea that promises to solve a scientific problem.[15] The first stage consists in racking one's brains by successive sallies of the imagination, while the second, which may be delayed for hours after one has ceased one's efforts, is the spontaneous appearance of the idea one has struggled for. Poincaré says that this spontaneous process consists in the integration of some of the material mobilized by thrusts of the imagination; he also tells us that these thrusts would be useless but for the fact that they are guided by the special anticipatory gifts of the scientist.

It seems plausible to assume, then, that two functions of the mind are jointly at work from the beginning to the end of an inquiry. One is the deliberately active powers of the imagination; the other is a spontaneous process of integration which we may call *intuition*. It is intuition that senses the presence of hidden resources for solving a problem and that launches the imagination in its pursuit. It is also intuition that forms our surmises in the course of this pursuit and eventually selects from the material mobilized by the imagination the relevant pieces of evidence and integrates them into the solution of the problem.

But what about the measurements, computations, and algebraic formulas used in the pursuits and formulations of science? These are, of course, essential; but just as the use of language is a tacit operation, both in our own speaking of language and in our understanding of what has been spoken by others, so it is also with all other explicit thought—with measurements, for example. Measurements can be developed and understood only by a tacit operation. They are based throughout on tacit knowing and are literally meaningless without it. All knowledge is therefore either tacit or rooted in tacit knowing.

We should now be able to see that all our knowledge is inescapably indeterminate. First of all, as we have seen, the bearing that empirical knowledge has upon reality is unspecifiable. There is nothing in any concept that points *objectively* or automatically to any sort of reality. That a concept relates to a reality is established only by a tacit judgment grounded in personal commitments, and we are unable to specify all these personal commitments or to show how they bring a given concept to bear upon reality and thus enable us to trust it as knowledge. We are unable to do this because we are *dwelling in* these basic commitments and are unable to focus our attention upon them without destroying their subsidiary function. The coherence that we see when we dwell in the particulars that make it up we see focally. Its particulars (the clues we dwell in) we see only subsidiarily. When we focus upon them, making them explicit entities (as in the example of the cat), we change their phenomenal character, and we find that they do not, in their new guise, logically imply—i.e., imply explicitly or in a detached manner—the reality that we do find them to imply through an indwelling *tacit* inference.

For these same reasons also, again as we have seen, our rules for establishing true coherences—as against illusory ones—are and must remain indeterminate. Any rules we have must be applied, of course; and, to do this, we may have additional rules for their application. But we cannot go on having specific rules for the application of specific rules for the application of specific rules ad infinitum. At some point we must have "rules" of application (if we can call them that) which we cannot specify, because we must simply dwell in them in a subsidiary way. They are part of our deepest commitments. But for this reason they are not specifiable.

Therefore, we cannot ultimately specify the grounds (either metaphysical or logical or empirical) upon which we hold that our knowledge is true. Being committed to such grounds, dwelling in them, we are projecting ourselves *to* what we believe to be true *from* or *through* these grounds. We cannot therefore see what they are. We cannot look *at* them since we are looking *with* them. They therefore must remain indeterminate.

The very process of tacit integration, which we have found so ubiquitous, is, when we turn our attention directly upon it, as we have been attempting to do in this work, also indeterminate, unspecifiable. We cannot spell this process out in explicit steps, and it is for this reason, as we have noted, that no "thinking" machine can ever be adequate as a substitute, or even as a model, for the human mind. Our dwelling in the particulars, the subsidiary clues, results in their synthesis into a focal object only by means of an act of our imagination—a leap of a logical gap; this does not come about by means of specifiable, explicit, logically operative steps. The depth *seen* through a stereoscope is a new phenomenal experience, not *deducible* in its unique phenomenological character from the clues that the process of tacit integration integrates, just as the heliocentric concept of the planets "seen" by Copernicus was a new conceptual experience *not* deducible from his available data. We can only point to the existence of tacit integration in our experience. We must be forever unable to give it an explicit specification.

But there is another indeterminacy, one not so clearly involved in what we have so far been discussing but nevertheless important to an adequate view of knowledge and science. When we modify our judgments about anything, we make subsidiary use of certain new principles—which is to say, we dwell in them. Because of this circumstance we actually make existential changes in ourselves when we modify our judgments. For we literally dwell in different principles from the ones we have been at home in, and we thus change the character of our lives. But these changes are not fully specifiable. First of all, since we are dwelling in these new principles in a subsidiary way, we are not able to focus our attention upon them and to render them explicit. Thus it is truer to say that we modify our grounds in making new judgments than that we explicitly modify our grounds and then make new judgments through their deliberate and explicit use. The new judgments first appear more meaningful to us than our old ones did, and so we commit ourselves to them—and thus indeterminately to the new principles also that ground them.

Second, even if we could know just what these new principles are to which we are committing ourselves, the consequences in our lives of our adopting them must remain indeterminate; for we cannot apply them to the situations we shall meet before we do apply them. These applications will each be an act of our creative imagination pulling ourselves and our situation at that time into some sort of meaningful integration. We cannot make these "discoveries" before we do in fact make them.

We have thus shown that the processes of knowing (and so also of science) in no way resemble an impersonal achievement of detached objectivity. They are rooted throughout (from our selection of a problem to the verification of a discovery) in personal acts of tacit integration. They are not grounded on explicit operations of logic. Scientific inquiry is accordingly a dynamic exercise of the imagination and is rooted in commitments and beliefs about the nature of things. It is a fiduciary act. It is far from any skepticism in itself. It depends upon firm beliefs. Nor should it ever give rise to skepticism. Its ideal is the discovery of coherence and meaning in that which we believe exists; it is not the reduction of everything to a meaningless jumble of atoms or accidentally achieved equilibrations of forces. Science is not thus the simon-pure, crystal-clear fount of all reliable knowledge and coherence, as it has for so long been presumed to be. Its method is not that of *detachment* but rather that of *involvement*. It rests, no less than our other ways of achieving meaning, upon various commitments which we personally share. We make use of these in science in creative and imaginative ways involving our very person. As we have seen, some share of indwelling is essential even to the meanings we synthesize in mathematics and physics. But in order to understand living things, we must dwell in our subjects of knowledge more deeply—and more deeply yet as, at each step, we seek to know higher animals, until we try to understand the highest animal of all, man. We can succeed here only by a completely reflexive indwelling—a full conviviality with our subject.

Thus the ideal of pure objectivity in knowing and in science has been shown to be a myth. It is perhaps a harmless myth if most of its implications are not followed out, but it is certainly a poisonous one if they are. For the implication that the truth about human behavior demands an amoral standpoint is, as we have seen, part of what our moral inversions have been made of. The other part, our current moral perfectionism, is something we have yet to investigate. But first let us draw an important conclusion from what we have been discussing in these last two chapters.

The understanding of science we have achieved in the preceding chapter enables us to see that the study of man in humanistic terms is not unscientific, since *all* meaningful integrations (including those achieved in science) exhibit a triadic structure consisting of the subsidiary, the focal, and the person, and all are thus inescapably *personal*. This observation, we noted, can be understood to constitute the first step in bridging the gulf that supposedly separates scientific from humanistic knowledge, attitudes, and methods. In view of what we have now seen in this chapter we can surely bridge this gulf completely. We now see that not only do the scientific and the humanistic both involve personal participation; we see that both also involve an active use of the imagination. That the various humanities are heavily entangled with the imagination has always been very clear to almost everyone; but that imagination has an essential role to play in science as well has rarely even been glimpsed. Science has been supposed by most modern thinkers to be a matter of ascertaining the hard, objective "facts." Sometimes, it was admitted, scientific inquiries are adventitiously beclouded by the working of some scientist's imagination, but only (it was to be hoped) temporarily so. Science has most commonly been thought to deal with facts, the humanities with values. But since, in this frame of reference, values must be totally different from facts, the humanities have been thought to deal only with fancies. Values have thus come to be understood to be the product of fancy, not of facts, and so not any part of factual knowledge.

This view turned out to be a basis for tacitly assuming that only the meanings achievable in science (the presumed sole discoverer of objective facts) could lay claim to be meanings having reference to realities—if any meanings could ever do so. The meanings achievable in the humanities lacked, by their very nature, any possible reference to reality. They were simply works of the imagination, brilliant in some cases, often sparkling and interesting, intriguing, and enjoyable, but nevertheless inescapably only ephemeral flashes of lights that never were—or could be—on land or sea.

The logic of this conclusion does seem, in fact, to be quite inescapable *if* one assumes that science, in its perfect state, is imagination-free—a work of pure detachment and objectivity, oriented only and solely by the facts. That perennial modern philosophic chestnut, the effort to derive values from facts, would seem to be as futile, therefore, as the ridiculous efforts to square the circle or to build a perpetual-motion machine. It might be suspected, however, that it is kept alive because of a strongly felt need to consider our humanistic meanings somehow as more than effervescent and willful fancies of our imagination.

If, however, as these chapters try to show, personal participation and imagination are *essentially* involved in science as well as in the humanities, meanings created in the sciences stand in no more favored relation to reality than do meanings created in the arts, in moral judgments, and in religion. At least they can stand in no more favored relation to reality on a basis of the supposed presence or absence of personal participation and imagination in the one rather than in the other. To have, or to refer to, reality—in some sense—may then be a possibility for both sorts of meanings, since the dichotomy between facts and values no longer seems to be a real distinction upon which to hang any conclusion.

But some distinction there must surely be between these two sorts of meanings. Let us therefore make a fresh analysis of the various sorts of humanistic meanings in terms of the triad we have seen to be at work in the meanings achieved in ordinary perception and in science. The real distinction between scientific and humanistic meanings must surely come to light in such an analysis. More important than the perception of this difference, however, must be the possibility of our becoming cognizant of the realities established in the world by the humanistic meanings that such an analysis may open up for us, as we come to see the grounds for the validity of such meanings and the vast importance they possess for us through the significance they are able to create in our lives.

4

FROM PERCEPTION
TO METAPHOR

MAN LIVES IN THE MEANINGS HE IS ABLE TO DISCERN. HE EXTENDS himself into that which he finds coherent and is at home there. These meanings can be of many kinds and sorts. Men believe in the reality of these meanings whenever they perceive them—unless some intellectual myth in which they also come to believe denies reality to some of them. Men are then in difficulty with themselves about these particular meanings. Since we call "real" any meaningful entity that we expect to manifest itself in unexpected ways in the future, we think of it as something that has a "life" of its own, so to speak. It is therefore not something we think of as a mere appearance, made up of the coincidental effects of many heterogeneous causes and subject entirely to the independent future manifestations of these separate causes—such as we believe the constellations of the stars to be. The stars in a constellation *appear*, from our position in the universe, to form a group; but we believe that the group will never manifest itself as a group in other unexpected ways, since our astronomical theories (in which we believe) tell us that these stars do not form an interacting, dynamic system.

Many of the coherences that we see around us we may not believe in, therefore, if our theories tell us they are not real. We may believe that they are only appearances, illusions, created by the chance interactions of many separate causes. Religious meanings may be reduced by us in this way to peculiar congeries of psychological needs and historical causes; ethical meanings to congeries of historically influenced economic needs; and aesthetic meanings to congeries of biological—or sometimes psychological—needs.

Although Bishop Butler's common sense told him that everything is what it is and not another thing, our experiences with illusions breed a common sense that knows also that everything is *not* always what it "is" (i.e., not always what it *seems* to us to be). So we cannot avoid the necessity of resorting to a personal judgment in order to decide when something is what it is and not another thing. And, as we have seen, our personal judgment is what *it* is because of the clues we dwell in, including, of course, the general views to which we are committed about the nature of things and the nature of knowledge. We ought, therefore, to adopt the kind of general views about the nature of things and the nature of knowledge that will not prevent our belief in the reality of those coherences that we do, in fact, see.

We have seen how we may adopt nonobstructive views with respect to perceptions and to the various coherences in nature that are open to our genuine modes of scientific inquiry. But coherences that are thought by us to be artificial, not natural, have had a difficult time being regarded and respected as real in our times, since (1) they seem to be creations of our own, not subject to the external checks of nature—and therefore to be wholly creatures of our own subjective whims and desires—and (2) only tangible things are supposed to be real.

We have seen, however, that such things as classes and minds, for instance, can be accepted by us as real even though they are not tangible. Classes and minds, however, are not understood to be creations of men. The meanings—the coherent entities—which we know as Michelangelo's *Moses*, Beethoven's Ninth Symphony, the virtue of justice, and the Christian God are not only intangibles; they are regarded by contemporary men as free human creations—not subject to correction by nature. They seem, possibly, to have no existence or being at all in the absence of man. Therefore, they would appear to have an existence only in the sense of being present, as such, in somebody's mind and in the sense of being the effect of a heterogeneity of natural causes. They may *appear* to us to be great and worthy of respect. But what if we suppose that they are only adventitious results of lower motivations or, eventually, of the reactions of atoms? What if we suppose, for example, that justice, although it purports to be a sort of universal fairness, is really only that which serves the interest of the ruling and owning class? If we suppose its *real* meaning, its basis in reality, is the interest of the stronger, as

Thrasymachus put it to Socrates so long ago, can we then respect it? Any would-be reductionist who could answer, "Yes, we can, if we can nevertheless see that it *is* fairness for all—if its meaning can be seen to be exactly that," is asserting that a coherence that he can in fact see has the status in reality that he seems to see it has. But he should then also admit that it has a status that cannot be tortured into a reduction to certain other causes or conditions of less import than itself, simply because in such a reduction the meaning that he can see it has would disappear. He should then simply give up his reductionist theory. He ought to be able to see that in fact he *has* given it up in this case. If, however, his scientistic attitudes are so deeply embedded in his mind that he cannot give them up, he may try to live in two worlds. He may believe (1) that the creations of men that he values so highly really are reducible to causes and to states of being that are lower than those they *appear* to him to possess. But he may also believe (2) that there are causes that generate in his mind the *appearance* that these meanings have a higher worth, and so he may try to go on treating these meanings with respect. But he can hardly help recognizing, at least tacitly, that he is then really allowing himself to be duped when he so honors them—that he is permitting himself willingly to suffer from illusions of their grandeur. The danger is that such a man may then also be induced at some point, as we saw in our first chapter, to steel himself to sacrifice these meanings to "the realities"—to the ruthless demands of a Marxism or a Fascism or to the no less ruthless demands of some other currently more fashionable theory of social engineering; for he knows no philosophic position that supports their reality other than in terms of those lower elements to which they are supposedly reducible.

In order to hold these meanings securely in the reverence they seem to him to demand, contemporary man therefore needs a theory of these meanings that explains how their coherence is no less real than the perceptual and scientific coherences he so readily accepts. He needs to see how his obvious personal involvement with these meanings is necessarily and legitimately part and parcel of the reality they actually have, that his personal involvement is not at all a reason to regard them as mere subjective fantasies. These meanings will then not seem to be mere appearances to him. They will seem to be in truth what they "are." That is, he will believe they are what they do honestly seem to him to be.

It would appear, therefore, that we need to extend our epistemology to those coherences that are often described as "artificial" as opposed to "natural." Let us begin the construction of such an extension by endeavoring to account for the kind of coherence and power to move us that exists in a metaphor.

The most elementary use of language is found in the use of a name to designate things like a particular person or building, and the simplest way of explaining how a particular name becomes attached to a particular person or object is to assume that this is the result of hearing the word spoken in the sight of the person or building. We are told a word in the presence of the object, and the coincidence of the two experiences becomes associated in our mind in the way associations are commonly formed between two things often seen or heard together. This explanation of the meaning attached to a name is so plausible that it has been widely accepted ever since the laws of association became known in the eighteenth century.

But, as we have seen, a word has a *meaning*. It *bears on* something else which is its meaning. A word and its object are not equal partners in an association. The explanation of language along associationist lines is thus fundamentally wrong.

Equal partners in an association are in fact easy to distinguish from the relation of names to their object. On entering Trafalgar Square in London, you can see the National Gallery and the Nelson Column. Once you have looked at them in turn, each might recall the other in an equal manner. But suppose you become aware of a tourist guide pointing at the Nelson Column; you notice the Column and the guide's finger in two different ways. The Column is interesting in itself, but the guide's finger is interesting only in its capacity for directing attention to something other than itself. If the guide then tells his audience the name of the Column, its members are not interested in the sound he utters but in this sound's capacity to direct their attention to something other than itself, i.e., the Column. They may remember both the Column and its name, but the Column will be remembered for its own sake, while its name will be remembered only because of its meaning, which is the Column. The word in use has in fact no interest in itself, as an object; in this it is very different from the object it names, which is interesting in itself as an object.

This refutes the theory of verbal meaning as an equal association of word and object and confirms instead our view that such meaning consists in a from-to relation. This conclusion has been anticipated to some extent by authoritative writers. The classical conception of language as a creative work of the mind does not envisage a purely associative relation of word and object. Edward Sapir, still writing in the spirit of this philosophy, emphatically rejects the idea that association may constitute speech. Bertrand Russell has observed the peculiar "transparency" of language, which is an aspect of the speaker's subsidiary awareness of words in contrast to his focal awareness of that which his words mean. Erwin Strauss and Susanne Langer have spoken in their different ways of the modesty of the sign in relation to the matters that it designates. These reflections lend support to our view of the from-to structure of language used in designating objects.[1]

Words, understood in this way, function as indicators, pointing in a subsidiary way *to* that focal integration upon which they bear. Some words can therefore be replaced by road signs, telling the way, or by maps or drawings by engineers, or by mathematical formulas, which also serve to help us find our way about things. These signs, maps, and formulas serve subsidiarily as indications, as denotative words do; and they have it in common with these words that, when they are viewed in themselves (not as they appear to us when they are serving their function of bearing on something else), there is little interest to be found in them. We can lump all these subsidiaries together as indicators pointing at something that is of intrinsic interest and recognize that, by contrast, *they* lack this intrinsic interest. And we can diagram this relation in this way:

We can see the subsidiaries (S) as bearing upon ($\longrightarrow$) their focal meaning (F).

$$S \longrightarrow F$$

We can then note that, in cases of indication, the subsidiaries (signs, such as words, maps, or mathematical formulas) are functionally of no intrinsic interest, while that upon which they bear is the part of the operation that claims our intrinsic interest (ii). This gives us a whole class of tacit-knowing operations that can conveniently be diagrammed as:

$$\begin{array}{cc} -ii & +ii \\ S & \longrightarrow F \end{array}$$

In these cases, which include practically all the kinds of coherences we have so far been talking about (integrations of perception and of conceptual knowing), the subsidiary clues are not of intrinsic interest in the transaction (– *ii*). It is the object of focal attention that possesses the intrinsic interest (+ *ii*). It is what is at the end of the cane that engages the blind man's interest, not the feelings in the palm of his hand. It is the meaning of a communication in words that engages our attention and interest, not the words as such. In fact, an accomplished linguist may not even be able to say later in what language a particularly interesting communication came to him!

These integrations might also be called self-centered integrations, because they are made *from* the self as a center (which includes all the subsidiary clues in which we dwell) *to* the object of our focal attention. Let us list here, in the nature of a summary reminder, twelve kinds of integrations that are self-centered in this sense in which the subsidiary clues lack the intrinsic interest that the focal object possesses:

Kinds of Self-centered Integration

Sensory clues fused to perception
Two retinal images fused to three-dimensional sight
Two stereo pictures fused to three-dimensional sight
Deliberate motions fused to intended performance
Actions taken in causing something to happen
Establishment of part-whole relations
Structure of a complex entity, e.g., a physiology
Series of integrations forming a stratification
Use of clues to establish reality of a discovery
A simulation identified with a simulated object
Recognition of a member of a class
Use of a name to designate an object

Let us now move into a consideration of meanings in which the subsidiary clues do *not* function (as they do above) merely as indicators pointing our way to something else. In this second kind of meaning it is the subsidiary clues that are of intrinsic interest to us, and they enter into meanings in such a way that we are *carried away* by these meanings. Our persons are involved in a way quite different from the way they are involved in self-centered integrations. Their involvement is of such a nature that the relation of "bearing upon" and the location of intrinsic interest become much more complex.

Let us first look at a sort of meaning in which the subsidiary

clues—those *from* which we attend focally to things—are of great
intrinsic interest, whereas that *to* which we attend focally is of little or
no intrinsic interest. Suppose we look at a flag, or a medal, or the
tombstone of a great man. As objects, these things have substantially
no interest to us; but what functions subsidiarily in bearing on, say, a
flag *is* of great intrinsic interest to us, for it includes our total
awareness of our membership in a nation. When we look at our
country's flag on a solemn occasion, this otherwise meaningless piece
of cloth becomes for us a moving spectacle and to some people even a
sacred object. Recall how linguistic from-to relations are similar to the
integration of parts to a whole. A name becomes attached to its object
to some extent and comes to form part of it. There is a similar link
between a nation and its solemnly unfolded flag: the nation's
existence, our diffuse and boundless memories of it and of our life in
it, become embodied in the flag—become part of it. The structure of
meaning found in medals, tombstones, and other things of this kind is
quite the same. Such intrinsically uninteresting objects of our focal
attention do not *indicate* something, as other intrinsically uninteresting things do, for example sounds used subsidiarily as words for
denoting an interesting object. Flags and tombstones *denote* a country
or a great man but they do not bear upon them as words bear upon
their objects; they rather *stand for* such interesting objects, which is to
say they *symbolize* them.

We can picture symbolization in the following way to show its
contrast to indication with respect to intrinsic interest:

$$+ \, ii \qquad - \, ii$$
$$S \longrightarrow F$$

The positions of the plus and minus signs have to be reversed in
symbolization because the subsidiary clues are more interesting to us
than the focal object. The focal object in symbolization, in contrast to
the focal object in indication, is of interest to us only because of its
symbolic connection with the subsidiary clues through which it
became a focal object. What bears upon the flag, as a word bears upon
its meaning, is the integration of our whole existence as lived in our
country. But this means that the meaning of the flag (the object of our
focal attention) is what it is because we have put our whole existence
into it. We have surrendered ourselves into that "piece of cloth"
(which would be all that the flag could be perceived to be were we to

try to view it in the *indication* way of recognizing meaning). It is only by virtue of our surrender to it that this piece of cloth *becomes* a flag and therefore becomes a symbol of our country.

Some of the subsidiaries, then, that bear upon the flag and give it meaning are our nation's existence and our diffuse and boundless memories of our life in it. These, however, not only bear upon the flag as other subsidiary clues bear upon their focal objects, but they also, in our surrender to the flag, become *embodied in it*. The flag thus reflects back upon its subsidiaries, fusing our diffuse memories. We cannot use a straight arrow to express this feature in our diagram, since such an arrow pictures only a straight, one-directional bearing-upon. We must make the arrow loop, in symbolization, in order to express the way our perception of the focal object also carries us back toward (and so provides us with a perceptual embodiment of) those diffuse memories of our own lives (i.e., of *ourselves*) which bore upon the focal object to begin with. This is how the symbol can be said to "carry us away." In surrendering ourselves, we, as selves, are picked up into the meaning of the symbol. We must therefore redraw our diagram something like this:

The integration of $\qquad + ii \qquad - ii$
our existence: $\qquad S \frown F$

This somersaulting arrow indicates the embodiment of ourselves—our existence as lived in our country—in the flag. In the surrender of ourselves to the flag, the medal, the tombstone—to whatever turns our focal object into a symbol for our country, for a great deed, or for a loved person—we accomplish the integration of those diffuse parts of ourselves that are related to these persons or things. Our surrender to these symbols is thus at the same time our being carried away by them. For this attainment of coherence does not take place in the serial manner in which we have had to discuss it: first this, then that, etc. It is a whole of which the various aspects of it that we have discussed are the functionally interrelated parts. These parts, or moments, operate at once in terms of one another. Our surrender and our being carried away are thus two sides of the same coin and occur at the same instant. We do not surrender to a symbol if we are not carried away by it, and we are not carried away by it if we do not surrender to it. It is a wholistic imaginative achievement of meaning, not a serialized, mechanical one.

Symbolization therefore entails something quite different from designation or indication. To designate the United States by its name is structurally the very opposite of symbolizing the United States by a flag. To designate the United States is to integrate a name to a country, while to symbolize the United States by a flag is to integrate a country to a flag. It is true that the *symbols* we have been considering are all inarticulate things (cloth, pieces of metal, stone), while the *indicators* we spoke of were words; but this difference does not account for the difference in the way they behave. When we pass on to poetry and later to myth and ritual, we shall see that the same difference will continue to exist between indication and symbolization even when both are accomplished through words. We need to form the concept of a class of meanings that covers all *artificial* cases of meaning as distinct from the meanings of perception, skills, and such part-whole relations as we meet in nature. Let us therefore use the word "semantic" for all kinds of artificial meanings—i.e., for all those that are contrived by man. The usual use of the term "semantics" limits it to the meanings achieved by language; we are expanding the term to include all meanings contrived by man. We will *not* then call "semantic" such kindred but substantially different meanings as those achieved by our trained powers of perception or by our productive skills. These will simply be called perceptions, or objects or actions achieved by our productive skills.

We have then, so far, two types of semantic meanings: indication and symbolization, which are inverse to each other with respect to the location of intrinsic interest. We shall shortly reach the principal subject of this chapter, another type of semantic meaning called "metaphor," where we shall see that *both* the S and the F may possess intrinsic interest. For now, however, let us note that the essential difference between indication and the whole group of meanings of which symbolization is one kind lies in the relation of the self to the whole process. Personal participation and indwelling of clues, though they are certainly always involved in all types of indications, tend, in indication, to integrate these clues into entities that seem to be projected away from the self as a center. Perception, for instance is of things seen from the self as a center. The self is never carried away in indication; it is never surrendered or given to the focal object. As we have noted, indications are always self-*centered*. By contrast, symbol-izations are self-*giving*. That is, the symbol, as an object of our focal awareness, is not merely established by an integration of subsidiary clues directed *from* the self to a focal object; it is also established by surrendering the diffuse memories and experiences of the self *into* this object, thus giving them a visible embodiment. This visible embodiment serves as a focal point for the integration of these diffuse aspects of the self into a felt unity, a tacit grasp of ourselves as a whole person, in spite of the manifold incompatibilities existing in our lives as lived. Instead of being a self-*centered* integration, a symbol becomes rather a self-*giving* one, an integration in which not only the symbol becomes integrated but the self also becomes integrated as it is carried away by the symbol—or given *to* it.

We are now ready to move into an understanding of metaphor. We find the field already well plowed, since many thinkers have tried to explain metaphor. Aristotle long ago noted that "It is a great thing, indeed, [for the poet to be able] to make a proper use of these poetical forms, as also of compounds and strange words. But the greatest thing by far is to be a master of metaphor." "Metaphor," he says, "consists in giving the thing a name that belongs to something else."[2] Owen Barfield in our day echoes this when he writes that metaphor is "saying one thing and meaning another."[3] Just why "misnaming" something in this way could move us so greatly has, however, been left ·unexplained. Aristotle may have hinted at a reason when he said that "a good metaphor implies an intuitive grasp of the similarity in dissimilars."[4] The pleasure that he thought all men take in recognizing something might seem to be involved in the interest we take in metaphors. Yet this general sort of pleasure would hardly seem to be of sufficient intensity or specificity to account for the power of metaphors to move us as they do.

Perhaps I. A. Richards has made as great an effort to explain the power of metaphors as anyone so far. It is both the likeness and the unlikeness of the two parts of a metaphor, the tenor and the vehicle, that somehow account, he thinks, for the way it works. There is, he says, a "peculiar modification of the tenor which the vehicle brings about," and this is "even more the work of their unlikenesses than of their likenesses."[5] But Richards does not explain to us how this peculiar interaction of likeness and unlikeness can have such a powerful effect upon us.

Except for very simple metaphors, which Max Black thinks can be viewed as substitutions or comparisons, an "interaction" view, such as he thinks I. A. Richards' view is, seems to Black to be the best. Yet there are many complications in this view. In the end Black seems to think that "the secret and mystery of metaphor" reside in the connection that *the reader* is forced to make between the two ideas in a metaphor, but how this works remains unexplained. In spite of the fact that he can and does use metaphors in his efforts to account for the "secret" of the metaphor (i.e., he has a tacit understanding of "metaphor"), Black fails to unravel this secret explicitly. In fact, he seems to think that, even if we could state a number of relevant relations between the tenor and the vehicle, "the set of literal statements so obtained will not have the same power to inform and enlighten as the original." Although he admits that an attempt to explicate the "grounds" of a metaphor can be valuable, he holds that we must not regard this "as an adequate cognitive substitute for the original." Thus, although Max Black seems to regard metaphors as communicating cognitive content, they seem to defy all his efforts to state just what this cognitive content is. A "suitable reader," he says, must "educe for himself, with a nice feeling for their relative priorities and degrees of importance," the various relations in the metaphor which, when we try to express them explicitly, we can only present wrongly as having equal weight.[6]

What is altogether missing in this honest admission by Black of his theory's limited ability to unpack metaphors is any sort of explanation of why or how a metaphor can *move* us so greatly—can carry us away. At best he shows only that we learn something cognitively from a metaphor that we did not know before and that this has something to do with a "suitable" reader's capacity to make a connection between "the two ideas" in a metaphor.

Let us take off from here and explore further this capacity readers may have to connect the diverse matters in a metaphor into a whole. It should be no stranger to us. We have met it in every instance of the integration of clues into a focal whole—integrations ranging from ordinary perception to generalization and including all our use and understanding of language. We have seen that most spoken sentences are unprecedented and hence are, strictly speaking, new creations; yet they are usually understood immediately. They present a problem that is easily solved. Indeed, so skillful is our interpretative machinery that, provided a sentence is formed grammatically, it is difficult to fill it with words, however absurd, that will not make some sense if one tries hard enough to interpret the sentence. Linguists regard the sentence "Colorless green ideas sleep furiously" as being fairly safe against making sense to anybody. But it is easy to see that "green ideas sleep furiously" may mean simply that "immature ideas foster violent dreams," and with a little trouble one should be able to account also for such ideas being colorless. The idea is not new. Leonard Bloomfield, a leading linguist in his time, suggested that absurd combinations of words will always find a poetic interpretation.

Man's well-nigh unlimited capacity to interpret grammatically formed sentences offers an opening for the literary incoherence that bitterly protests the state of man in our day. André Breton, who declared such an intention for his every thought and action, claimed that, "to compare two objects, as remote from one another in character as possible, or by any other method put them together in a striking and sudden fashion, this remains the highest task to which poetry can aspire."[7] Ezra Pound's lines "In a Station of the Metro" offer an example:

> The apparition of these faces in the crowd,
> Petals on a wet, black bough.[8]

This expression of fragmentation, which refuses to accord any meaning to our modern world, will reappear throughout these chapters. It is spoken of by Yeats in three lines which also give us an example of a cool, intellectual use of metaphor:

> Shakespearean fish swam the sea, far away from land,
> Romantic fish swam in nets coming to the hand;
> What are all those fish that lie gasping on the strand?[9]

To declare a passionate faith in the mercy of Christ crucified, T. S. Eliot composed these intricately involved metaphoric lines:

> The wounded surgeon plies the steel
> That questions the distempered part ...;[10]

And the full-throated voice of passion speaks in Shakespeare's metaphor in which Richard II defies the enemies who are conspiring to depose him:

> Not all the waters of the rough rude sea
> Can wash the balm from off an anointed king.[11]

The structure of metaphor and the source of its powers can be demonstrated briefly in this last example. Translated quite literally into plain prose, it would say that the anointing balm sticks so firmly to a king who has been anointed that all the waters of the sea cannot wash it away—with the implication that an anointed king cannot be deprived of his office because he cannot be deprived of his balm![12] Such a claim is absurd because the hypothesis upon which it rests, if taken naturalistically, is ridiculous; but the claim advanced in this metaphor, understood *metaphorically*, is clear and forceful; it is not an absurd claim about a mere physical eventuality. It does indeed affirm such a claim in the literal meaning of the words used, but it means something more than this literal meaning. The semantic mechanism by which a clear and forceful metaphorical meaning is established is the same as that by which a flag is made to symbolize a country—with the difference that the flag (as a piece of cloth) is meaningless in itself, while the verbal projection of the seas trying in vain to wash the balm from a king, though fanciful, is far from meaningless. In fact it presents a tremendous spectacle to our imagination.

We may thus see that when a symbol embodying a significant matter has a significance of its own and this is akin to the matter that it embodies, the result is a metaphor.

Since both the tenor and the vehicle in a metaphor have intrinsic interest—both are significant ideas or expressions in themselves—we can diagram a metaphor thus:

$$
\begin{array}{cc}
t & v \\
+\,ii & +\,ii \\
S \overset{\frown}{} \circ \overset{\frown}{} F
\end{array}
$$

The tenor bears on the vehicle, but, as in the case of a symbol, the vehicle (the focal object) returns back to the tenor (the subsidiary element) and enhances its meaning, so that the tenor, in addition to bearing on, also becomes embodied in the vehicle.

We can now schematize the way our rapture in a metaphor arises by adding a level involving ourselves, thus:

$$
\begin{array}{ccc}
\text{Ourselves} & & (t \;\overset{\frown}{\circ}\; v) \\
+\,ii & & +\,ii \quad +\,ii \\
S & \overset{\frown}{\bigcirc} \searrow & F
\end{array}
$$

As in the symbol, so in the metaphor: the subsidiary clues—

consisting of all those inchoate experiences in our own lives that are related to the two parts of a metaphor—are integrated into the meaning of a tenor and a vehicle as they are related to each other in a focal object (a metaphor). The result is that a metaphor, like a symbol, carries us away, embodies us in itself, and moves us deeply as we surrender ourselves to it.

The metaphor from *Richard II*, the story of the sea and the balm, which in a literal sense is preposterous, is suffused with feelings—with the king's angry pride and defiance—and so becomes enlarged into a powerful and moving image, embodying our own diffuse experiences and thus giving us an object in which to see them as integrated.

In other metaphors, for example in the two lines of Ezra Pound, the integrated meaning of two matters may be more equally supplied by both than was the case in the sea-and-balm metaphor, and in the lines of Eliot we have what Empson describes as a virtually mutual metaphor.[13] A metaphor may be passionate, like the one spoken by Richard II, or it can be impassioned by very different feelings, as in Eliot's two lines. It may sparkle by elucidating an interesting subject, as the three lines of Yeats do. But all these variations can be seen to be covered by man's basic imaginative capacity for integrating two or more disparate matters into a single novel meaning. In the next chapter this point will be illustrated further with respect to works of art.

It is commonly known that metaphors, like jokes, lose their effectiveness if they are explained in detail. We know that a semantic integration is destroyed if we switch our attention from its meaning to that of which it is the meaning—in other words, from the point of focal attention to the subsidiaries which bear on that focus. Sometimes a shattered semantic integration can be replaced, even profitably replaced, by an explicit relationship; but in many cases this is impossible, as the lost power of an explicated metaphor shows. We have seen how foolish Richard II would sound if he announced the inviolability of his kingship by pointing out that this can be compared with the incapacity of all the raging seas of the world to wash off his royal unction—perhaps because his balm possessed the property of hygroscopicity.

As a rule a metaphor loses its force even when transposed into a simile. But similes can be immensely powerful. The power of these lines by Baudelaire is unsurpassed:[14]

Le Poëte est semblable au prince des nuées,
Qui hante la tempête et se rit de l'archer.

On the other hand, Dom Moraes has told in his biography how, in his
student days, he showed Auden a poem with the line "Women
swaying their long hair, like trees." "That won't do," Auden said
crossly, "It won't do at all. You can have trees swaying their long hair,
or women swaying their long hair, but one swaying its long hair like
the other won't do."[15] The metaphor or the straight statement was
acceptable, but the simile was intolerably prosy. The more subtle
factors of meaning deprivation cannot be strictly defined.

The effects of rhythm and rhyme and other formal features of
poetry can be explained on lines akin to the interpretation of
metaphor. The two constituent parts of a metaphor are made to bear
on a joint novel meaning of them. We are aware of them subsidiarily
in their joint focal appearance. This seems to hold also for the formal
features of a poem. In reading a poem we are aware subsidiarily of its
rhythm, its rhymes, its sounds, its grammatical construction, and the
peculiar connotations of the words used. Each of these components
can be examined separately, in itself, but this inevitably dims and may
even efface the meaning of the poem. Its meaning may be brought
back to us with a deeper understanding when we turn our focal
attention back upon the poem instead of upon its parts; on the other
hand, the poem may prove to have irrevocably lost some of its
freshness. In any case, our awareness of the components, which we
have focally examined, must once more become subsidiary if we are to
see the poem's meaning.

In other words, the rhythm, rhyme, sound, grammar, and all the
other more subtle formal aspects of a poem, along with the several
allusions of its parts, all jointly bear on the meaning of the poem. We
are not therefore aware focally of what they add to that meaning and
how they affect its quality.

This rich and delicate pattern of subsidiaries imbues a poem with
the quality of a distinctive artifact. It lends the poem harmonies that
no other speech possesses and establishes its claim to be received for its
own sake. It sets poetry off, detaches it from the ordinary run of life. A
poem's story is thus exempted from being heard as a mere communication of facts and asks to be heard instead by the imagination.
Therein lies its independence as a work of art. I. A. Richards long ago

described the isolating effect of poetic form, which we are interpreting here by the principles of semantic integration. He wrote:

> Through its very appearance of artificiality metre produces in the highest degree the "frame" effect, isolating the poetic experience from the accidents and irrelevancies of everyday existence . . . Much which in prose would be too personal or too insistent, which might awaken irrelevant conjectures or might "overstep itself" is managed without disaster in verse.[16]

One may wonder indeed how the content of Shakespeare's eighteenth sonnet, "Shall I compare thee to a summer's day?", would sound in prose. After some compliments to the beauty of his mistress, nearly half the sonnet is spent telling her that all her beauty will pass away. She is then suddenly reassured, most emphatically, that she will be an exception to this fate, only to be told next that this wonderful promise means only that the poet's genius will keep this sonnet—and herself—famous forever. This story, which in prose sounds shabby, is redeemed by the beauties of the sonnet. We will look at this power of poetry in the next chapter, in the wider perspective of all the arts.

5

WORKS OF ART

WE HAVE SEEN THAT AN EXPLICIT ACCOUNT OF A SHAKESPEAREAN metaphor turned it into grandiloquent nonsense and, similarly, that the prose rendering of one of his sonnets reduced its content to a piece of callous self-adulation.

Further, we have seen that our new theory of meaning presents us with the basic mechanism of the way specific explication brings about such destruction. The mechanism for destroying the meaning of a word consists in shifting our focal attention from the meaning of the word to the word as a perceptual object. Likewise, the meaning of a metaphor is destroyed when we shift our focal attention from its meaning to the constituent parts of the metaphor, and the meaning of a poem is destroyed when we turn our focal attention from the poem's meaning to the prosaic or "story" content of the poem, which is only one of the subsidiaries bearing upon the poem.

But the damage done to metaphors and poems by specification includes a loss that is much more noticeable than the loss that is incurred when we break off the bearing of a word on its object by focusing our attention upon the word itself. The subsidiaries composing metaphors and poems are joined together by an imaginative performance much richer than any imaginative action required for linking a word to its meaning. To reduce a metaphor or a poem to its disconnected subsidiaries is to extinguish the vision which linked them to their integrated meaning in a metaphor or poem. What is left is but a caricature of their true meaning. This is, we hold, why Max Black found he could not, without loss of meaning, explicitly specify all the "relations" between the two elements in a metaphor.

To arrive at the structure of poetry we must, however, distinguish it from metaphor. To integrate two disparate matters does not amount to a work of art, although such integrations may constitute some of the *parts* of a work of art. Maybe the twentieth century will reject the very idea of works of art before it comes to a close; but a rejection of art would only add precision to the conception of art. What is being rejected would at length become more clearly known through the process of rejecting it, just as our own inquiry into the loss of meaning in our time must elucidate the meaning of meaning before we can speak of its loss.

We have moved from talking about poetry to speaking of works of art in general, and this extension was deliberate. Our study of works of art will include painting, sculpture, and stagecraft, in addition to poetry. We shall then have before us all the arts that clearly represent something, and we shall limit our discussion in this chapter to these, the representative arts.

Let us start from the old question of why works of art continue to be honored as "true" even though they tell us stories that we clearly understand are not true. I. A. Richards gives us a good example of this paradox. He notes that the theatrical presentation of a murder "has a different effect upon us from that which would be produced by . . . actual murders if they took place before us."[1] The question is just what this different effect is and how it is produced. Let us recall a similar paradox generated by a metaphor: the paradox that in a metaphor we say one thing and mean something else. We have a similar case in the theater. In witnessing a murder on the stage, we are aware of the setting and the antecedents of the stage murder, which are incompatible with the murder's being genuine; yet—just as in the case of the metaphor—we do not reject these contradictory affirmations, which would make the stage murder a nonsensical deception, but call upon our imaginative powers to integrate incompatible matters into a joint meaning. This joint meaning has, in a play, the peculiar quality of a dramatic event visible only to the imagination, just as the meaning of a metaphor, produced by integration of its two incompatible constituents, is known to us only in our imagination.

A dramatic event is a work of art, and the structure of dramatic events is typical of all works of art. But stage plays have an exceptional range of subsidiaries. The playwright, the director, the actors, the designers, the whole theater, and the mechanism of stage properties are all involved in a dramatic performance. Poetry and painting are comparatively simpler cases of the structure of art.

By analogy to the murder in the theater, to which we do not respond as if it were real, we could say that the effect of Shakespeare's sonnet differs altogether from the effect of its content when stated in prose, because the meaning of the sonnet is rooted in a host of poetic subsidiaries which are disregarded in the prose account of the sonnet's content. But there is more to this. The sonnet as a work of art is not merely enriched and altogether recast by its poetic subsidiaries; these subsidiaries also serve to *cut the sonnet off from the person of the poet*.

In order to see this, we have to correct an oversimplification we have until now been making in our account of integration. We have talked about the integration of retinal snapshots and other subsidiary clues into a perception, of a set of movements into a skill, of the integration of what are, strictly speaking, disparate objects into the formation of a comprehensive class that covers them; we then went on to deal with the integration of widely different ideas into a metaphor, and this led us at last to poetic integration. We have given full weight to the fact that, as we ascend from physiologically performed integrations to the formation of concepts, metaphors, and, finally, works of art, integration requires an increasing measure of imaginative effort. Since men notoriously differ in their imaginative faculties, it could be inferred that many would be unable to perform an integration which others, more gifted for this task, may achieve.

We could add to this the obvious fact that the very possibility of a highly imaginative integration will remain undiscovered until it is actually carried out by a mind of exceptional powers. This may have been clear enough as an implication of our account of the whole range of integrations. What was never made clear, however, was that there are limits to the possibility of integrating things. In nature, untouched by man, all things may be said to hang together to some extent; but we commonly recognize sets of distinct objects that cohere to each other far more firmly than they do when we try to attach them to other distinct objects, such as fire and smoke, clouds and rain. Modern science has discovered a vast network of coherences unknown in previous ages and has dissolved as illusory many seeming coherences that before were accepted as genuine, such as constellations of stars. Modern engineering has invented innumerable new coherences in the shape of contrivances. Poetry, painting, and drama have populated our imagination with works coherent in a number of ways. Thus, works of science, engineering, and the arts are *all* achieved by the imagination. However, once a scientist has made a discovery or an engineer has produced a new mechanism, the possession of these things by others requires little effort of the imagination. *This is not the case in the arts.* The capacity of a creative artist's imaginative vision may be enormous, but it is only the vision that he imparts to his public that enables his art to live for others. Thus the meanings he can create for his public are limited by the requirement that they provide a basis for their re-creation by the imaginations of other viewers or readers. The *use* of a work of art by others is not, therefore, like the use of an invention, such as the telephone. We do not have to recreate A. G. Bell's imaginative vision of the telephone in order to use it, nor do we have to do this in order to know and use Newton's laws. But we do have to achieve an imaginative vision in order to "use" a work of art, that is, to understand and enjoy it aesthetically.

It is, however, fortunate that our vision does not have to include the artist's strictly personal feelings, memories, etc. It does not have to because, whenever our powers of integration produce a coherence, they do so by cutting off the subsidiaries of this integrated body from connection with other experiences. This, in fact, is the principle which turns every discovery, invention, or work of art into a sort of reality with life, so to speak, of its own—into a body of living thought severed from continued dependence upon the personality of its maker and thus capable of being understood by other personalities.

In poetry this necessary detachment from a particular personality is furthered by the artificiality of the "frame" into which the poem is cast. I. A. Richards points this out to us: "Through its very appearance of artificiality metre produces in the highest degree the 'frame' effect, isolating the poetic experience from the accidents and irrelevancies of everyday existence."[2]

Rhythm is but one artificiality of a poem among many others. Rhymes, expressive sounds, and peculiar grammatical constructions, strange connotations of words, and, above all, metaphors are other poetic accessories. They all function as subsidiaries which, combined with such content of the poem as can be put into prose (i.e., its "story"), form the meaning of the poem. But people normally talk in prose, and this is true also for the usual conversations of poets. Anyone who addressed us in rhymed and metrical speech would be taken to be doing this as a joke, unless we suspected a pathological obsession, similar to that of obsessive punning. A poem can form no part of a poet's *usual* personality. Thus the formal structure of a poem, which has so much of the poem's meaning in it, forms a blockage, insulating the poem from everyday affairs and so also from the poet as a private person. When entranced by a poem, we repeat its words through our lifetime; strictly speaking, it is *the poem* that speaks to us, not the poet.

The same holds, of course, for plays. The persons of Hamlet, Othello, and Shylock are known in themselves and not as part of Shakespeare's person. A painting by Cézanne can be instantly recognized by many people who know next to nothing of the painter. These facts are so obvious that one hesitates to state them in such detail. But their significance is far from being fully recognized.

I have compared Shakespeare's eighteenth sonnet with the poet's telling his mistress, in ordinary conversation, "You are beautiful, but you will fade and die; however, your beauty will become immortal through my immortal verse." We can see now that the difference between them is twofold: (1) the poem is not the voice of the poet, and (2) its meaning is not conveyed by its prose content. For its meaning is formed by the integration of its formal pattern with that part of its content that can be expressed in prose. But what its prose content expresses is incompatible with the artificial speech of a poem. Nevertheless, we succeed in integrating these incompatibles—the artificial pattern and the prose content—and by doing this we produce a joint meaning which is the meaning of the poem. *Logically* speaking, however, these parts of the poem would still be quite incompatible. In imaginatively attaining this meaning we are subsidiarily aware of its two components, which, viewed in themselves, are explicitly or logically incompatible but which, when combined by an artistic imagination, speak in one voice as the subsidiaries of the poem's meaning. This is how the grotesqueness of its prose content is dissolved in the lines of Shakespeare's eighteenth sonnet.

We may therefore illustrate in the following diagram some part of the way in which the meaning of a poem or work of art is achieved:

$$\text{Work of art} = \text{frame} \underset{\circ}{\overset{\circ}{\subset\supset}} \text{story}$$

The frame and the story embody each other. Neither "bears on" the other nor symbolizes the other.

This integration of parts in the meaning of a work of art is but an instance of the rule—exemplified in many different ways in our present work—that the integration of subsidiaries produces a perception differing in both appearance and content from its constituents. What is produced by the *poetic* imagination, however, is a *radical* novelty, and its reader absorbs this novelty by the powers of his own imagination. Thus we, the readers, come to share in our enjoyment of a poem the detached position the poet has in his relation to his own poem. He has sent it forth into the world, detaching it even from his own daily affairs by giving it its artificial frame. We are able, therefore, to enjoy it *in itself*—not as we enjoy the satisfaction of our personal desires. This is how we can watch a murder in a play in integrated conjunction with its theatrical subsidiaries without either jumping up to rescue the victim or feeling the action on the stage—the pretense of a murder—to be nonsensical. We accept the clues which the play offers to the imagination for sharing its meaning, and we live in this meaning rather than the meaning these events would have for us in our ordinary "interested" lives.[3]

This is something of what Kant meant when he defined the aesthetic appreciation of art as a disinterested pleasure; it also accords with the claim made for painting by Conrad Fiedler, that art is a production of reality by a mind.[4] It is not the mere satisfaction of subjective, substantive desires. The work of art is a "something"—a "reality" with powers of its own. But these terms are worn out. They must be resharpened by carrying further our structural analysis of representative art.

From the vantage point of this analysis we can see that poems and also paintings, sculptures, and plays are so many closed packages of clues, portable and lasting. Their durability is infinitely superior to that of our personal experiences, for the coherence of their parts is so much firmer and more effectively organized. The contrast between the looseness of our lives and these organized "parcels" of art has been described effectively for the case of poetry by I. A. Richards:

> In ordinary life a thousand considerations prohibit for most
> of us any complete working out of our response; the
> range and complexity of the impulse-systems involved is less;

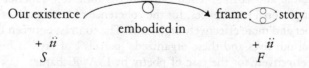

88 CHAPTER FIVE

the need for action, the comparative uncertainty and vagueness of the situation, the intrusion of accidental irrelevancies,
inconvenient temporal spacing—the action being too slow or
too fast—all these obscure the issue and prevent the full
development of the experience. We have to jump to some
rough and ready solution.[5]

Richards contrasts this condition with the severely circumscribed
existence of poetry as achieved and absorbed by its audience.

This transition is stated more definitely by T. S. Eliot. He writes:

When a poet's mind is perfectly equipped for his work, it
is constantly amalgamating disparate experiences; the ordinary man's experience is chaotic, irregular, fragmentary. The
latter falls in love, or reads Spinoza, and these two experiences have nothing to do with each other, or with the noise
of the typewriter or the smell of cooking; in the mind of the
poet these experiences are always forming new wholes.[6]

Something more than the integration of its frame and its story
therefore occurs in our grasp of the reality of a poem. The poem takes
us out of the diffuse existence of our ordinary life into something
clearly beyond this and draws from the great store of our inchoate
emotional experiences a circumscribed entity of passionate feelings.
First the artist produces from his own diffuse existence a shape
circumscribed in a brief space and a short time—a shape wholly
incommensurable with the substance of its origins. Then we respond
to this shape by surrendering from our own diffuse memories of
moving events a gift of purely resonant feelings. The total experience
is of a wholly novel entity, an imaginative integration of incompatibles
on all sides.

We must therefore add another somersaulting arrow to our previous
diagram of what goes on in an integration in poetry or art:

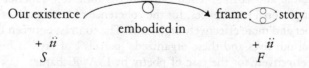

Our existence   frame   story
embodied in
+ *ii*   + *ii*
S   F

But let us set aside for the moment any further discussion of the
powers that poetry can exercise over our minds so that we can see how
other kinds of art, especially painting, also fit into this pattern.

Let us think of painting as it was done in Europe for close to a thousand years—from the time of Pericles to the Byzantine period, and then again from the time of Giotto and Duccio up to the end of the nineteenth century. This kind of painting aimed, at least according to expressed views, at making a likeness of things perceived or imagined. Giotto was hailed for making painting lifelike. A century later the imitation of depth by central perspective was developed to the general applause of the Italian public. The last major innovation that promised closer simulation of nature was that of French Impressionism. Then only there came a change: the twentieth century opened with a challenge to the ideal of simulation. We mention this challenge now only in passing.

The strange thing is that, all through these centuries, works of painting and sculpture were poured out by artists to the general acclaim of their public. That is, there was no feeling of disappointment that these works never really produced the illusion that an intended simulation should surely aim at, namely, that it be *mistaken for* the things which it represented. Such verisimilitude was hardly ever achieved; and when upon rare occasions it was, it was not hailed as the final triumph of visual art. Pictorial illusions of this kind were known, rather, as oddities.

In the chapter on "Ambiguities of the Third Dimension" in his *Art and Illusion*, E. H. Gombrich surveys past observations on the manifest contradiction between simulated depth in paintings and the actual flatness of the canvas. He quotes the French neoclassical critic Quartremère de Quincy:

> "When the painter packs a vast expanse into a narrow space, when he leads me across the depths of the infinite on a flat surface, and makes the air circulate . . ., I love to abandon myself to his illusions, but I want the frame to be there, I want to know that what I see is actually nothing but a canvas or a simple plane."[7]

Gombrich points out, however, that psychologists have found that illusions are destroyed when we shift our attention to an alternative view which contradicts the illusion. "Is it possible," he asks, "to 'see' both the plane surface and the battle horse at the same time?" He replies that this is impossible: "To understand the battle horse is for a moment to disregard the plane surface. We cannot have it both ways."

It is true, of course, that alternative views representing alternative integrations of the same scene may totally exclude each other. But we have also seen ample evidence of the existence of elements that contradict each other when seen focally but which can nevertheless, as subsidiaries, be integrated into a joint meaning. This seems to be the case here. The conflict between the portrayal of depth in the painting and the flatness of its ground is resolved by the same mechanism which fuses the contradictory parts of a metaphor. This same mechanism also fuses the formal pattern of a poem with its content (that part of it that can be given in prose). Thus the painting as a work of art is unified in terms of visual qualities that do not exist in its separate parts.

Let us refer here to some experimental work by psychologists that shows how sensory contradictions are resolved in terms of a sensory innovation. Irving Rock and Charles S. Harris have demonstrated that such sensory resolutions of sensory contradictions are actually made. A subject who is required to doodle while wearing a right-left inverting prism soon feels his hand to be at the place where his eyes, wrongly, show it to be: asked to write down some letters and figures, he writes them right-left instead of left-right. A more far-reaching integration of conflicting clues, in terms of sensory innovations, has been found to underlie the way one finds one's way about when wearing inverting spectacles. The work of Kottenhoff, as we may recall, has shown that the subject learns to find his way by reintegrating the inverted images to all his proprioceptive, auditory, and gravitational clues.[8] Sensory innovations of this kind amalgamate conflicting clues just as the viewing of a painting unites the conflicting clues of depth and flatness into the total meaning of the painting.

The theory that our subsidiary awareness of the canvas affects the way we see a picture was first put forward by M. H. Pirenne. He pondered the strange behavior of the picture covering the ceiling of the Church of Saint Ignatius in Rome, done by the Jesuit Andrea Pozzo at about the turn of the seventeenth century. This painting shows, among other figures, a set of columns which appear to be continuations of the pilasters supporting the ceiling. This effect, however, can be seen only from the middle of the aisle; from other parts of the floor the columns appear curved and lying back at angles to the supporting structure of the church. Pirenne met these facts by suggesting that the viewing of an ordinary painting from the side does not produce a distortion of appearance; distortion is produced only when the perspectival illusion of the painting is so perfect that it makes the painting appear *really* three-dimensional. All other sorts of paintings are insensitive to the angle of viewing; since their perspectival design is not fully convincing, the viewer remains aware that he is facing a flat canvas. Pirenne believes that this awareness of the flat canvas reduces the illusionary powers of the perspective and thus enables us to retain, at skew angles of viewing, an undistorted appearance of the painting. He adds that the particular appearance we choose is the one we see from the correct center of perspective. Our eyes seem to select this appearance for all positions of viewing, because it resembles the way the painted object is seen in nature.[9]

When Pirenne describes how, in looking at a painting, we are aware of its flat canvas, he says that we have a "subsidiary awareness" of the canvas—using this term in the same sense in which we have used it in our theory of semantic integration. I gratefully acknowledge that it was Pirenne's work that initiated my treatment of the way we view a painting. His argument is not reproduced here, however, since I would prefer not to rely altogether on the particular evidence he uses.

We can now distinguish sharply three ways in which we can be aware of brushstrokes and canvas in a painting: (1) in a *trompe l'oeil* work the brushstrokes are integrated to a joint meaning which is identical with that of the external object represented; (2) in a *work of art* the brushstrokes are integrated to a joint meaning with background elements like the canvas, of which we are totally unaware in viewing a *trompe l'oeil* and which are indeed focally incompatible with the external object represented by the painting; and (3) in a *focal view of the brushstrokes and the canvas* their semantic integration is destroyed and the painting is reduced to a meaningless aggregate of paint blots covering a flat surface.

We have yet to explain why an illusory likeness of its object was not acclaimed as the perfect achievement of simulation during the long period in which artists strove in all ways to perfect simulation. The answer may be found, as we have seen before, in the superiority of poetic expression over the effect of our personal experiences. When we see the external objects represented in a painting as though they were these actual objects, their significance is reduced to being some objects among innumerable others, which usually means that they will be seen from a trivial point of view. We realize instantly the abysmal triviality to which a still life by Cézanne would be reduced were it made to convey the perfect illusion of real fruits and vegetables offered for sale in a niche in an art-gallery wall. We can then appreciate the fact that the joint integration of brushstrokes and canvas lends to all paintings a distinctive artificial quality which distinguishes them from all natural sights. It secures the *artistic reality* of a painting and so guards its distinctive powers from dissolving into the surroundings of *factual reality*.

Sculptures are usually secure against being taken for the subject they represent. But suppose we became the victims of Madame Tussaud's skill. Suppose we addressed a hat-check girl and found her to be made of plaster. We would not hail this marvelously lifelike product of plastic simulation as a work of art surpassing Michelangelo's *Moses* on the grounds that the *Moses* is too easily recognized as a mere statue. The same thing applies to the stage. An actor playing Hamlet who created the illusion that he was *actually* dying from the poisoned rapier of Laertes would produce a disgusting disturbance. His aim must be to produce the play in which Hamlet is killed by Laertes.

Hence Coleridge's view that art requires a "willing suspension of unbelief" is doubly misleading. We do not appreciate a work of art, whether it be poetry, painting, sculpture, or drama, by suspending our disbelief in its prose content. A work of art represents certain facts of the imagination. It does not affirm any fact of *experience*. If we believed a work of art to be simply affirming certain facts of experience, we would not see it as a work of art. It could at best be seen as a clever illusion—worthless to us as art. Appreciation of a work of art requires belief in what it *means*, not unbelief in something it does not and must not *assert*.

We recognize the meaning of a work of art without special effort if the style of the work is familiar to us. Styles have changed continuously since the Periclean age of Greece, but the changes were slow enough to be understood and welcomed by later times. However, the last hundred years have brought innovations in all forms of art so radical and frequent that they have continuously *challenged* hitherto accepted styles and standards. Most of these innovations were therefore violently opposed for some time; but many of them soon gained acceptance, and this eventually resulted in a complete transformation of most of the arts—in particular poetry, painting, sculpture, and drama, the representative arts which we have dealt with here. As we shall see, these changes have brought out rather clearly a number of facts about art not clearly known before.

Initially, opposition to these innovations denied to the artists who created them the support they needed for their subsistence. These artists accepted the sacrifice required, however, and, renouncing the comforts of bourgeois society, they founded a community accepting neediness within a permissive, subversive bohemianism. The profound dedication of these innovators was matched by the fierceness of the opposition ranged against them. Upholders of the academic tradition were not satisfied with ignoring the works of art which they did not appreciate. They went on to threaten the livelihood of the innovators, denouncing them as frauds and sometimes violently interrupting the performance of modern plays and operas. They showed an attachment to traditional forms of art as strong as that of the innovators to their own new ideas.

This situation gave rise eventually to a group of critics who recognized the modernists' artistic worth and supported them by trying to explain their principles. This support in turn evoked numerous disciples, who became convinced that the new forms of art held meanings which were worth the trouble of discovery. At last it became part of a modern education to explore by exercise of the imagination the meaning of these new and puzzling works of art.

These conflicts lasted from about the mid-nineteenth century to the thirties of this century. We had two opposing camps, bourgeois academics and bohemian moderns, the latter supported by avant-garde critics and that section of the public that strives to keep up with supposed artistic progress.

In a later chapter we shall deal in more detail with the changes in artistic content that took place during this period. For now let us take note only of the strange transformation these changes produced in the relation between the artist and his public. Trained during years of battling for new forms of art, modernist critics and their followers developed an unprecedented facility for interpreting unusual art forms, and this facility spread widely among the public. This reduced resistance to artistic innovations to such a degree that innovators could henceforth count on speedy recognition. The battle between bourgeoisie and bohemians was ended. The result was a changed relation between artists and their public. Poets, painters, sculptors, and playwrights, as well as novelists and film producers, could count on the capacity and willingness of the public to develop slight and often esoteric clues in the light of their own imagination.

These new art forms and the new criticism that explained them also brought out clearly the fact that we can grasp a work of art only through the efforts of our imagination—whether the work of art is representative or not.

6

VALIDITY IN ART

THE LIFE OF ART IN SOCIETY, AS WE HAVE SEEN, IS THE WORK OF THE artist's imagination renewed by the imagination of those who receive it. We have seen this imagination rooted in the act of perception and in every deliberate motion of our limbs. Even scientific discovery can be taken to include an imaginative perception of things hitherto unknown. The technical structure of a bird's flight or of man's upright walk presents, when analyzed, an ingenious combination of bodily parts functioning together like the parts of a machine. These are achievements of meaning in themselves, and, moreover, they are the primordial groundwork for the invention of skills, the use of tools, and the facts of engineering. It is worth noting that to arrive at the structure of art we have moved from the physiological achievements of meaning to the structures of language, conceptual terms, and symbols and metaphors and that along this way we have met works that call for an increasing amount of imaginative activity in order to account for them. We have avoided until now the question, on what ground we may justify or evaluate poems or paintings or plays. To get at this question, the center of all past aesthetics, let us follow a line parallel to the line of the development of imaginative meanings that we have been following—a line leading from perception and voluntary motoric actions to science and technology—comparing the way valid meaning is achieved in science and technology with the way it is achieved in works of art.

Science and technical invention can best be understood if we watch them as they initiate an inquiry and guide its pursuit. Let us therefore review what we have said about these matters in regard to science and then see how they operate also in technical invention and art.

We have noted that scientific inquiry consists of three parts: first the finding of a problem, then an inquiry into the problem, and finally, if the search is successful, the solving of the problem. In recognizing a good problem and deciding to pursue it, the scientist must show an exceptional sensitivity to promising clues. He must be able to guess with a high degree of probability that there is something important lying hidden in a particular direction, and he must be able to assess the effort and expense of the inquiry. In a modest way all of us commonly exercise such faculties, but the scientist must be exceptionally gifted with them if his inquiry is not to be doomed before it starts.

All three parts of a scientific inquiry are set in motion by two mental powers. They receive their guidance from integrative powers, while they are propelled, and also supplied with suitable material, by thrusts of the imagination. The integrative powers are largely spontaneous; to mark this, we may give them the name of "intuition." All the labor and anguish involved in the creative process go into the thrusts of the imagination; intuition is effortless. At the inception of an inquiry, intuition predominates. Imagination enters at this stage only by keeping intuition alert to the sensing of a problem. We may describe these anticipatory judgments that guide the sighting of a problem and the decision to inquire into it as the work of a "strategic intuition."

An inquiry opens by a thrust of the imagination in the general direction suggested by the problem. The thrust, if successful, will reduce the vagueness of the problem and offer a firmer guidance for the next push toward a possible solution. The whole course of the quest is filled by laborious efforts of the imagination, broadly guided by a questing intuition, which also continues to select from the fragments mobilized by the imagination those which promise to become part of the solution.

Poincaré, who has described this process, has told how the quest is often brought to a close after a quiet interval (when the efforts of the imagination are at rest) by a sudden illumination which offers a solution for the problem. Such an event is purely spontaneous and so may be called the work of a "concluding intuition." The result may turn out to be false, and we may find ourselves sent back to resume our quest—and even, perhaps, to fail in the end. But there is always the same story over again. First, an idea appears, given by intuition, to be pondered by the imagination. Second, the imagination is let loose to ferret out a path of possible clues, guided by intuitive feelings. And third, an idea offers itself intuitively as a possible conclusion, to be pondered in its turn in the light of the imagination.

Technical inventions are made through the same three stages. One sees a problem, anticipates its feasibility and worthwhileness, thrusts one's imagination in a direction that promises success, and finally sights a solution that appears satisfactory. However, the content differs from that of science. The aim of a scientific inquiry is indeterminate, and much of the pursuit of what we dimly anticipate is indeterminate too. The inventor's aim, on the other hand, is relatively fixed, the way the aim of a deliberate movement is or the effort to acquire a skill. And, of course, in technology the test of an apparent solution lies in a more practical direction than it does in scientific discoveries.

It is important to note, however, that throughout the middle section of any scientific inquiry the imagination is heavily engaged in its quest for the missing solution. In this it must be guided by powers of anticipation, since otherwise its chances of hitting on an appropriate hypothesis would be one in a million. This point is fundamental. The imagination does not work like a computer, surveying millions of possibly useless alternatives; rather it works by producing ideas that are guided by a fine sense of their plausibility, ideas which contain aspects of the solution from the start.

The imagination is still at work at the other end of a scientific quest too, when a solution has been found. The thrusts of the imagination have subsided, but a belief that we have found a solution to our problem is fraught with anticipations of further manifestations, which we can entertain only in the imagination. A new theory that claims to be real anticipates by this claim an indefinite range of future, as yet unknown, manifestations. The solution of a technical problem has perhaps less widely indeterminate implications, but they are wide enough to substantially engage the imagination.

This completes our preparation for comparing works of art with discoveries and inventions. Let us keep in mind the from-to structure of meaning. The subsidiaries bear on a focus; they mean something *to which* we attend *from them*. We have seen how such a semantic relation is established by integrating subsidiaries to a focus. This led us to the integration of incompatibles into works of art consisting of: (1)

artificial patterns—"frames"—and (2) "stories"—contents that can be stated in prose. We saw how our imagination produced such an object of the imagination from these incompatibles.

The quest for scientific discovery also integrates fragmentary clues to an initially unknown coherent meaning (although the quest is guided vaguely by certain powers of anticipation), whereas technical invention starts, on the contrary, by aiming at a product that will fulfill a definite function and seeking the means to contrive it. To produce a work of art is to make something never before seen but grasped in a vague way by powers of anticipation, and in this essential feature the artist's quest is nearer to that of the scientist than to that of the inventor. This may seem strange, for the artist does not find things hidden in nature, as the scientist does, but contrives his product as inventors do. In fact some arts, like painting, are mere variants of a handicraft. Yet H. W. Janson is right in saying that

> the making of a work of art has little in common with what we ordinarily mean by "making." It is a strange and risky business in which the maker never quite knows what he is making until he has actually made it; or to put it another way, it is a game of find-and-seek in which the seeker is not sure what he is looking for until he has found it.[1]

But the kinship of art to technology is important too. This becomes more clearly evident if we put these comparisons in terms of from-to integration. A scientific problem consists of subsidiaries anticipating an unknown focus. A technical problem consists of a desirable focus anticipating subsidiaries that will implement it. The scientist's quest has the structure of asking "What do these words mean?", while the engineer's quest has the structure of asking "What words will express my meaning?"

To search for words to express one's meaning may seem to be exactly what a poet does and hence to be, perhaps, central to all art. But remember that the *meaning* of a poem comes into existence *only with its words*. The poet starts with a problem that is largely indeterminate at both ends: it is open in its aim as much as in the means it uses for achieving it. One can watch this best in a painter's progress. H. W. Janson describes it:

> ... the creative process consists of a long series of leaps of the imagination and the artist's attempts to give them form by

shaping the material accordingly. The hand tries to carry out the commands of the imagination and hopefully puts down a brush stroke, but the result may not be quite what had been expected, partly because all matter resists the human will, partly because the image in the artist's mind is constantly shifting and changing, so that the commands of the imagination cannot be very precise.... In this way, by a constant flow of impulses back and forth between his mind and the partly shaped material before him, he gradually defines more and more of the image, until at last all of it has been given visible form.[2]

In the twentieth century we have seen certain developments in physics which have changed the very terms in which we understand nature, and the inventions of modern technology have frequently included the invention of the needs which they satisfied. Science and technology have, to this extent, brought changes in the very substance of the intellectual and material existence of man. But all the arts work in this way. They search for means of solving a problem—a problem which was conceived for this very purpose, i.e., its solution; and they pursue this quest while continuing to shape the problem so that it will better fit the means for solving it. The way a child grows to a higher level of physical maturity provides us with an illustration of how this kind of creative process, involving a continuous interaction of both means and ends, operates. But this is a work of nature. Art is the deliberate creative growth of man's existence.

As we have seen, the inception of a scientific inquiry and the undertaking of a technical problem are both based on imaginative anticipations of unknown facts, but to start on a work of art is to anticipate a result which will be brought into existence first in the imagination of the artist and then in that of his public. An artistic problem is the imaginative anticipation, not of unknown facts that already do exist, in some sense, in nature, but of a fact of the imagination—of a poem or a painting that *could* exist. We have noted that the artist's work is a constant invention of means for expressing his aims, coupled with readjustment of his aims in the light of his means. This manner of deliberate growth resembles scientific or technical inquiry in sometimes offering opportunities for sudden inspiration and at other times demanding the taking of infinite pains. Examples of both exist in every art.

Perhaps the most important difference between the arts, on the one hand, and science and technology, on the other, is found at the end of their pursuit, in the way the two are tested. Technical inventions and scientific discoveries are subjected to much more impersonal tests than works of art are. Yet this difference, though important, is not absolute; and we will understand the nature of artistic validity better if we show the way personal criteria apply even to the more impersonal works of the mind.

In chapters 2 and 3 we saw the extent to which personal participation is involved in science. In view of this—and the consequent indeterminacies that this participation entails—it should be evident that we have no strict proofs for any parts of science. The radical import of this fact is usually blunted by contentions that the statements of science are only probable and merely tentative; but this is an exaggeration and is in any case irrelevant. The fact is that not only do we accept and vitally rely on scientific observations (we do not really treat them as only probable or tentative), but we do so on the ground of *nonstrict* criteria. Our reliance on the validity of a scientific conclusion depends ultimately on a judgment of coherence; and as there can exist no strict criterion for coherence, our judgment of it must always remain a qualitative, nonformal, tacit, personal judgment.

The recognition of a proposed contribution to science is thus controlled by a system of subtle values which are decisive in shaping the very conception of science. There are actually three main scientific values. Other matters being equal, the more clearly and impersonally we can establish the variable on which a contribution to science relies, the more precious it is; and again, other things being equal, the deeper and wider the systematic aspects of the contribution, the more valuable it is; and finally, both these values are transcended by a prescientific value: the ordinary interest of the subject matter. The proportion in which these three values are found in different sciences varies greatly, yet these varying combinations must be assessed jointly as one total value, for this is essential for the rational conduct of science.

But who is to assess this value? How are these values to be cultivated without suppressing the originality which may demand a renewal of values? Once again, there are no strict criteria for making these decisions. The pursuit of science is obviously fraught with value judgments, and by doubts about how to exercise such value judgments, in many of the same ways as the pursuit of any art.

The scientist, applying nonstrict criteria to the evaluation of scientific merit, does so in the conviction that these criteria are universally valid, and the scientific opinion of the time endorses this claim. It requires such value judgments to be "objective" and relies on them to be so. Accordingly, their validity is attested to by the authority of scientists *as a body*—not simply by the authority of the personal judgment of the contributing scientist. The success of science in universally imposing such self-set standards of value lends support to a similar practice in the life of the arts. To see this kinship in its full depth, we must first take notice of those features of the arts in which it is most manifest.

The arts are works of the imagination, and so are the sciences. But all our hopes and fears, all our memories and our very feeling of ourselves, our suppressed desires and hidden feelings of remorse, all that we see in sleep and indeed in daytime perceptions, and all our deliberate bodily motions—all these are also works of the imagination. Why then does the word "imagination" instantly evoke in our minds the notion of works of art rather than any of these other matters? The reason that comes to mind with little effort is that the arts alone aim at transmitting their imagination to a public—to successive generations of publics—and depend on the imaginative powers of these people to accept the works of their imagination as meaningful. But we can explain also what it is that qualifies the arts—and the arts alone—for this enterprise. Our lives are formless, submerged in a hundred cross-currents. The arts are imaginative representations, hewn into artificial patterns; and these patterns, when jointly integrated with an important content, produce a meaning of distinctive quality. These artificial patterns are, as we have seen, what isolate works of art from the shapeless flow of both personal existence and public life. They make of works of art something detached, in many cases portable and reproducible, and potentially deathless.

In poetry, the intensity of artistic imagination is fed by the invention of far-reaching metaphors. The pursuit of science may also invoke powers of the imagination for integrating widely scattered pieces of evidence to form a new discovery; but the fusion of these previously unconnected clues will thereafter be quietly accepted as a fact. Not so the fusion of disparate subjects in a metaphor. The metaphor will continue to feed much of the imaginative fire that served to create it, and those who respond to the metaphor will continue to be moved by answering visions and feelings.

When the artificial frame of a work of art, integrated to its prose content, establishes a detached work of art, it also sets forth a claim that its value is universally valid. This is the point which we anticipated when we passed from science to the arts. The artist may rightly argue that when he claims universal validity for his self-set standards of value, he is but acting as scientists do when they claim objective validity for their own self-set standards of scientific value.

But the maker of a work of art claims more than this. A work of art bears the mark of its creator. Name a major painter or sculptor, poet or playwright, of past centuries, and his manner will readily come to the mind of many lovers of his art. A scientist will also possess his own style in setting up problems and pursuing them, and he may bring about some changes in the standards of scientific interest and even of validity. Once accepted, however, these modified standards will be applied in their own inquiries by all scientists. Only a forger will try to paint new Renoirs or Cézannes or to write new Shakespeare sonnets or Ibsen plays. By framing his work, the artist detaches his product from his personal life, but by this very act he includes *his own* unique artistic problem and his solution of it in the frame that demarcates the property he offers to the public. In fact, it is because of the acknowledged value of this embodiment in a unique frame that forgeries of a particular artist's work can be attempted and have value. No one would dream of forging a new discovery by Einstein. It is only the artist who detaches himself as an artist from himself as a private individual and embodies *this* artistic person in his work. Scientists cannot do this. But therefore all art is intensely personal *and* strictly detached; and it must, as we said, claim universal validity for the *personal* self-set standards which it obeys.

Such a claim clearly goes beyond the claim made for the sorts of meanings and discoveries that scientists accept as objectively established. That a certain scientist has created a theory is not regarded, even by him, as sufficient reason for anyone to accept and value it. But we must note that an artist's claim to autonomy does *not* substantially exceed the scientist's claim to autonomy based on the self-set standards which guide the pursuit of science. Mature scientists are trusted by their scientific community to use considerable sums of money—and the time of their assistants—to explore problems which they personally deem to be more promising than other possible lines of inquiry, and they are trusted still further in pursuing their own hunches. Their imaginative anticipations of feasibility and worth-whileness are accepted, and these assessments have proved right sufficiently often to justify this policy.

One might think that, on somewhat similar lines, one could shift the final responsibility for accepting artistic innovations to the public, since art cannot become a public possession unless the public is moved by it, and this might be thought to mean that the *whole* artistic community must include both artists and their public. But a little reflection will enable us to see that this would extend the range of relevant self-set artistic standards to include the self-set standards of nonartists. The fact that twentieth-century audiences have learned to appreciate almost any innovation might well make us hesitate to rely on their critical judgment. One would prefer to rely rather on those who must labor and suffer to satisfy their self-set standards by their own work. Their perseverance in the teeth of public rejection may often be a better test of self-set standards than a ready public acceptance of their work. This was certainly so in the period from 1870 to 1930.

Art has no tests external to art. Its making and acceptance must therefore be ultimately grounded *on the decision of its maker*, interacting, it is true, with both tradition and the public's present inclinations, but nevertheless interacting by and through the maker's own judgments. The fact that the artist must labor to meet his self-set standards is sufficient warrant that he submits to these as being universal standards, not of his own arbitrary or willful making. He may be the first ever to recognize them, yet he feels himself bound by them, not superior to them; for to him his innovation of standards appears to be a discovery, just as the innovative creation of a statistical understanding of nature appeared to modern physicists to be a discovery.

That these grounds of artistic creation are ultimate does not mean, however, that they are infallible; they may be contested by other artists, just as the statistical explanation of quantum theory was contested by other scientists. They may eventually be abandoned. But this would be a change (made by artists themselves) to other self-set standards, the adoption of which would then be the ultimate justification for all work done under their guidance. Actually these standards themselves would have been established, as all principles are, by the *valid work* done under their control. They would not have been chosen by a deliberate act (even a responsible one) *before* meaningful work had been done under their control. As we saw in chapter 3, new principles come into existence as part of the subsidiary clues establishing a new coherence. Only after this new coherence has been accomplished is it possible to see what the new principles are that ground it. This is why we can say that principles and standards are self-set but never deliberately chosen. Such is the structure of a responsible commitment as distinct from a purely subjective, arbitrary, fanciful "choice."

This structure applies equally to the arts and the sciences—and likewise to the public, whether in approving or rejecting works of art. But it is the public that must learn its criteria from the artists, not the artists from the public. Whenever we are faced with the necessity of deciding on a judgment, we cannot avoid relying on ultimate criteria. Even a failure to judge demands that we rely on some ultimate criteria for our refusal to judge. The point is, however, that we are often unaware of what these criteria are until after we have relied on them as subsidiary clues in a focal integration.

There still remains a difference between the ultimate criteria of the arts and the sciences. The merits of a statistical formulation of quantum theory, on the basis of which the theory becomes accepted, are speculative, whereas when an art is renewed—for example, when painting was revived by Impressionism—the attraction lies in the imaginative powers unleashed by the new vision. The same thing is true in judging the sort of world view projected by science. Such world views are properly judged, not by the speculative criteria appropriate to science, but by the criteria by which art is judged—by the attraction of the imaginative powers set free by this vision of the world; for the extension of scientific thinking into the formation of a world view is *a work of the imagination*, not of the formally critical intellect.

This is an important point, since the main influence of science on modern man has not been, as is often supposed, through the advancement of technology; it has come, rather, through the imaginative effects of science on our world view. The industrial revolution came about without substantial aid from the scientific discoveries made up to that time, but the imaginative effects of the Copernican revolution were already widespread. The visible universe had been immensely expanded, the earth thrown out of its supposed central location, and the ultimate grounds of man's existence reduced to the mechanics of matter in motion. During the past eighty years or so the progress of science has become a mainspring of technical progress, and this has changed many of our habits, improved our material welfare, and brought us certain special problems; but it has not had anything like the profound effect upon our conception of ourselves as human beings that Darwinism has had, and Darwinism has been responsible for no technical progress. It was not technology that produced the totalitarian ideologies which brought the disasters of the twentieth century into being, along with the feeling of absurdity and the contempt for human society that are current today. We may thank the scientific image of the world, as reflected in the modern mind, for these.

Such images cannot be tested by experience in the way the actual contents of science can be. These world-view images are works of the imagination which take parts of science as their subject. Like all works of fiction, they must treat their topic in a well-informed, plausible manner. Recall the famous statement by Laplace, that an intelligence that knew at one moment of time

> all the forces by which nature is animated and the respective situation of all the beings who compose it ... would embrace in the same formula the movements of the greatest bodies of the universe and those of the lightest atom; for it, nothing would be uncertain and the future, as the past, would be present to its eyes.[3]

This statement has the plausibility of a good passage of science fiction. Although we can show that it is absurd, the demonstration of its absurdity makes no impression on those scientists, or those non-scientists, who love its magnificent flourish, exalting a brilliant aspect of science.

This is not to decry works of the imagination presenting a world view based fundamentally on science. Far from it. These very pages are searching for the outline of such a world view. But at this moment we are concerned mainly with showing at what point the scientific imagination becomes based on the same principles that underlie the arts. Thus, when Laplace writes that "the future, as the past, would be present to its eyes," we ought to read this as we read Shakespeare's line "But thy eternal summer shall not fade"; and we should note that Laplace, and all those who draw images of the world inspired by science, enter that area of freedom and power which is properly that of poetry or painting, of plays, or of any of the other arts. It was perhaps not by accident that Lucretius was a poet or that Plato before him, in his *Timaeus*, labeled his own world view a "likely story."

We have developed I. A. Richards' theory that the arts can move us more deeply than ordinary "real" events can because they associate their thematic content with the artificial features of a frame. We have also endeavored to show that this association of content and frame is an integration that produces its own incomparable, purely imaginative experience—an experience that can be judged by no external criteria—and we would like to make such imaginative experience the corner-stone of aesthetic theory. To move man aesthetically is to move his imagination to make such integrations.

Aesthetics has spoken through the ages of the harmony and beauty that please us in the arts. But other beauty can also please us. The intellectual beauty of a scientific theory is pleasing, and so is the beauty of a sunset or a woman; and the word "beauty" is used today very freely to praise an ingenious invention, an elegant combination in chess, or a supreme feat of athletics. But these beauties hardly move our imagination, except in terms of special interests of a personal or professional kind. Beauty of this kind is really too harmonious for art, which depends for its self-assertion on bridging incompatible elements by the powers of its imaginative integration. Modern poetry, modern painting, and modern plays and stagecraft have proved this clearly. Only a considerable imaginative effort to achieve an integration of their parts can disclose a meaning for them.

But we have seen other floods of the imagination in our day in the terrible flights of modern ideologies, shaped by the scientific world view. Ethics and religion are also forceful expressions of the imagination and may become powerfully interwoven with works of art. But the only work of the imagination that has rivaled the power of art to move us has been the power of totalitarian ideologies.

The Soviet government decreed that dialectical materialism must be the world view of science. This imposition has had some effects upon some important branches of biological research and has also collided briefly with the modern principles of physics and physical chemistry. Taken over the first fifty years of its rule, the official philosophy has not, however, retarded the progress of *science* in the U.S.S.R. to any great degree. Scientific achievements do not appear to have been less distinguished during this period than they were in the previous half-century, when compared with other European nations. By contrast, however, the effect of the Soviet ideology on the *arts* was devastating. Scientists went on living in peace, but those who tried to develop their own imaginations in the arts were regarded as a menace to the state and were treated as such. In painting, where Russia's contribution was least distinguished before 1917, submission to the doctrine of "socialist realism" caused almost total destruction and produced a style on a level of empty philistinism that could find no place at all in Europe. In the other arts, where the position of Russia had been among the first in Europe, the Russian contribution fell back to near insignificance.

This shows at once how much the powers of the artistic imagination depend upon the autonomy of the artists and also how much power lies in the essentially artistic imagination when it creates a world view. The pursuit of science, as such, involves the imagination only minimally in comparison with the pursuit of the arts. In fact, the pursuit of science is of real importance only in its bearing on man's thought, feeling, and purpose when, on the basis of science, the artistic imagination develops a so-called scientific world view. But then its bearing on our lives appears to be all-pervasive. For this reason we must judge the quality of a "scientific" world view by the richness of its imaginative integrations, just as artists judge the quality of their own productions in art.

We must not forget, therefore, that our over-all task in this inquiry is not simply to develop, on a sound epistemological basis, the semantics of the artificial coherences · discovered (or created) by man—as if this were a cool, professional, academic task—but to produce, in a manner akin to art, a new moving vision of the world, imaginatively richer in the scope of its integration of disparate parts than those we have heretofore been offered by our scientific myth-makers.

7

VISIONARY ART

THE OXFORD DICTIONARY OF QUOTATIONS LISTS ABOUT SEVENTY thousand key words referring to about the same number of lines, mainly of poetry. This shows the wide range of English phrases, coined by poets, that English-speaking people can use in shaping their thoughts, and this is only a fragment of the ideas that are available to us. We borrow ideas from other poetry than English, as well as from plays, paintings, and sculptures. It is fashionable today to scorn borrowed ideas. We are told that they cannot express genuine convictions. But this is absurd. As we have seen, we cannot start discovering new ideas, even in science, without first adopting a whole framework of ideas which others have had before us. Indeed, the very idea of demanding absolute autonomy for our thoughts is itself a traditional doctrine. It goes back at least to Descartes, and its modern radical form, teaching us to choose our own values, is Nietzsche's doctrine and is now itself nearly a hundred years old.

Among the ideas readily available in our culture, a tremendous number are transmitted to us in imaginative form and are accepted by us on such terms. We can understand how this happens if we think of art as an extension of perception. We are accustomed to regard our perception of objects as our response *to them*; but, as we saw in chapter 2, we may instead regard perception as an act of tacit inference aiming at a correct interpretation of the traces made in our body by external objects. Understanding perception in this way, as an integration of experience, we may understand the artist's vision also as an integration of experience which, like perception, can succeed or fail—or can achieve partial success. The artist himself will judge in the first place whether his product is a true response to his experience. Then his readers or viewers will respond to his meaning by the experience the artist's work evokes in them and will accept it as true if they are deeply moved by it.

We have seen that, for this to happen, the artist's interpretation of experience must differ sharply from our usual perceptions. It must not represent daily experiences and current problems. It must see its subject, as it were, in one moment, so that its message can be transmitted in the equally brief moment of its presentation. This means, as we have also seen, that the creative individuality of the artist becomes distinct from the vagaries of his own entirely private existence. Setting a work of art apart from the concrete practical problems of some unique person's daily life gives it a universal and yet personal significance to which others may also be able to respond.

This condensation and insulation of art is achieved by its artificial pattern, which infuses the work of art with peculiar qualities that distinguish it sharply from our ordinary experiences. It is in terms of such artistic expressions that poetry, painting, sculpture, and drama convey to succeeding generations the whole range of ideas of which the current use of seventy thousand lines of English poetry is a token. The artist condenses his understanding of the things men have seen and done; and when this understanding appeals to us, we make it our own and clarify our lives by it. Art moves us, therefore, through influencing the lived quality of our very existence. In other words, without art our existence would mean much less to us. Some works of art continue to do this for centuries. Their meanings are of sufficient depth and universality to embody those basic aspects of human experience that remain unchanged through time. This is what is meant by calling great art "immortal."

We showed in chapter 5 that the acceptance of a work of art by its public is not due to a suspension of disbelief (as Coleridge held) but is due rather to our immersion in the belief that works of art are meaningful; sustained by this belief, we then eventually succeed in discovering a joint meaning for the focally incompatible elements in the particular work of art in front of us. This refutation of the notion that our acceptance of art is due to our suspension of disbelief will be given its full weight if we go back to the antecedents of the fallacy in question. We find what is perhaps its first statement in Book Ten of Plato's *Republic* (595a–599e), where Socrates calls "mimesis," i.e., the simulation supposedly practiced by painters, a falsification of the truth. Plato's concern about the deceptive nature of art may have been generated by two developments of Greek thought in the fifth century B.C. Greek art had taken a revolutionary step toward naturalism. Gombrich has called it a change from "making" to "matching," that is, from making symbols to simulating objects.[1] At the same time, philosophy developed a predominant desire to distinguish sharply between appearance and reality. The imperfection of the new imitative art was thus open to objection in this light. But the problem was more general and entered into the heritage of Western critical thought in the form in which Coleridge was eventually to meet it.

The romantic movement of the nineteenth century mitigated the dilemma by claiming that the content of art is predominantly subjective, personal. Thus it does not imitate. It merely expresses our subjectively personal feelings. But the progressive sharpening of skeptical thought, leading to the wholesale questioning of all traditional values, including the value of the individual person, espoused by the romantic movement, was presently to make any emphatic statements of man's deeper feelings sound trivial and to make their expression in conventional artistic patterns sound false. Plato's critique of art, showing that artistic mimesis is a falsification of the truth, thus reappeared in our day to support an attempt to make art abandon all explicit expression of positive content. An explicit expression of our inner states was considered too trivial; an explicit expression of the truth about things, impossible.

Today this movement has been going on for more than a century. It started in poetry as far back as Baudelaire's *Fleurs du mal*, published in 1857. It spread to painting and sculpture in various kinds of Postimpressionism and then to the stage, with Pirandello, in the period between the wars, and it has been extended to fiction, in the form of the antinovel, during the past twenty years.

In his preface to the *Fleurs du mal* Théophile Gautier wrote that Baudelaire had banished eloquence and passion from poetry. Such a banishment is even conveyed by Baudelaire's dedication of his volume to the reader, whom he addresses as "—Hypocrite lecteur,—mon semblable,—mon frère!"[2] The refinements of poetic language, cultivated for centuries, are brushed aside, and a sequence of hardly coherent images, with sordid admixtures, is thrown into the face of a society believed by the poet to have been corrupted by its own lies. The unique imaginative powers generated by juxtaposing words that are incoherent at first sight was developed much further by Rimbaud about fifteen years later. His poem "Bateau ivre" established definitively the new conception of poetry, first approximated by Baudelaire.

Listen to a translation of the fifth of twenty-five quatrains composing this poem:

> Sweeter than the bite of sour apples to a child,
> The green water seeped through my wooden hull,
> Rinsed me of blue wine stains and vomit,
> Broke apart grappling iron and rudder.

And in the original French:[3]

> Plus douce qu'aux enfants la chaire des pommes sures,
> L'eau verte pénétra ma coque de sapin
> Et des taches de vins bleues et des vomissures
> Me lava, dispersant gouvernail et grapin.

The sound of the French original is tempting to the ear, and the sequence of images, now twisted, now flowing, alternately surprises and rewards our attention. Trained as we are after a hundred years, during which this poem has inspired numberless successors, we have no great difficulty in restraining ourselves from asking for an explanation of the prose content of these verses. It was the exercise of this restraint that Rimbaud had in mind when he told a friend that a poet must make "himself a *seer* by a long, gigantic and rational *derangement* of *all the senses*."[4] When he said that the poet must be a seer, he meant that he must be capable of fusing a primarily incoherent sequence of images into a powerful joint meaning.

The technique of Baudelaire and Rimbaud was adopted by many poets who succeeded them. The stream of this influence is reflected in the attack Tolstoy made on it in his essay "What Is Art?", first published in 1897. He denounced Baudelaire, Verlaine, and Mallarmé for establishing a whole decadent school of meaningless poetry.[5] He could have indicted many more poets, especially in France, for belonging to this school.

It was not until England and America had suffered the shattering experience of the First World War that the influence of the new poetry reached them, fifty years after its foundation in Paris. Eliot's *Waste Land*, published in 1922, was among its firstfruits. When this poem was rejected by reviewers as unintelligible, I. A. Richards came to its rescue in an article that echoed Rimbaud's appeal that no point-by-point meaning should be sought in such poetry. He blamed Middleton Murray's inability to understand the poem upon his overintellectual approach. Appealing to those "who still give their feelings precedence to their thoughts, who can accept and unify an experience without trying to catch it in an intellectual net or to squeeze out a doctrine," he then went on to describe Eliot's technique as a music of ideas:

> The ideas are of all kinds, abstract and concrete, general and particular, and, like the musician's phrases, they are arranged, not that they may tell us something, but that their effects in us may combine into a coherent whole of feeling and attitude and produce a peculiar liberation of the will. They are there to be responded to, not to be pondered or worked out.[6]

The new poetry became known under various names. Three are widespread: one is symbolism, another is intuitionism, and a third is formalism. The name symbolism tells us that matters mentioned in these poems are not to be thought of in themselves, but metaphorically, as bearing on the poem's whole meaning. The term "intuitionism" refers to the fact that the integration of the poem's several parts to form a whole is essentially spontaneous; both the way the poet forms the poem and the way his reader grasps its meaning are said to be essentially spontaneous. And to call such poetry "formal" or "structural" tells us that the significance of its several phrases is muted, that it is the total structure of the poem that engages the imagination. But it seems to be more to the point to call such poetry "visionary," for its meaning is created by a powerful act of the imagination which comprehends all details in one. It must be grasped by a visionary experience, as the poet himself grasped it in his act of creation. Accordingly, in this present work we shall call such poetry, and all arts akin to it, "visionary."

Since a purely visionary poem says nothing that can be expressed in a prose statement, the problem of mimesis cannot arise. The visionary poem's affirmation is like that of nonrepresentative paintings, which show familiar figures in fragments or in absurd combinations, possibly distorted or fantastically colored; it asserts, as surrealist or cubist painters do, that its disparate elements have a joint meaning, a meaning that will be the more strikingly novel—more exciting and moving—the more incompatible its unintegrated elements are.

A renewal of painting, also in a rebellious style, coincided in time and in place with the rise of visionary poetry. It started as Impressionism, a coherent movement in the early sixties in Paris, and ran much of its course there before spreading over Continental Europe, from Russia to Italy, by the turn of the century. Its challenge to empty conventions was in tune with the new poetry, but its program was aimed chiefly in the opposite direction: not at emancipating the artist from giving a faithful account of experience but, on the contrary, insisting upon the production of what was thought to be a far truer representation of nature than that provided by the academics who ruled the taste of the age.

Whether it was actually more faithful to nature or not, Impressionism certainly included a novel exercise of the imagination and spread among the public a new habit (of which we spoke in the last chapter) of deliberately widening the range of its artistic vision. And I think it was this experience that opened the way, at the turn of the century, to the Italian Futurists, the French Fauves, the German Expressionists, the Russian Suprematists, and, eventually, to the various kinds of abstract painters, all of whom were united in emancipating, in their several ways, the painter's visionary powers from the simulation of nature.

We shall return in a moment to the social subversiveness which accompanied these artistic innovations. Suffice it to say now that this subversion was fiercely intensified by the experience of the First World War and the victory of the Communist revolution in Russia.

This was the moment when innovations in painting caught up with the art of visionary poetry. Initiated during the First World War in Switzerland by Dadaism, the new painting was effectively launched only a few years later in Paris by the program of Surrealism. In chapter 4, in connection with metaphors, we referred to André Breton's idea that "to bring together two objects, as remote from one another in character as possible, and unite them in a striking and sudden fashion is the highest task to which poetry can aspire." This declaration affiliates Surrealism to the school of visionary poets, and the practice of surrealist painting bears out this affiliation. Surrealist compositions are filled with utterly incoherent shapes and representations, and this brilliant absurdity is often reinforced by painting interwoven sections

in widely disparate perspectives. Surrealist painting is thus visionary
painting which belatedly joins visionary poetry as the second visionary
art.

The theater and the film were even later in following the principles
of visionary art. Ionesco's *Rhinoceros* and Beckett's *Waiting for Godot*
were pioneers on the stage; Robbe-Grillet's *Last Year in Marienbad*
was among the earliest "abstract" films to have a wide audience. And
the same author's novel, *Le Voyeur*, may be added as an early
representative of the antinovel.

These kinds of visionary arts are perhaps even more enigmatic than
their precursors in poetry and painting; for plays, films, and novels
commonly speak to us in the language of ordinary communications,
and we expect them to grip our attention without any effort on our
part. We therefore find their visionary form unintelligible until we
realize that we must not try to understand them as representing a
sequence of events that hang together in the way real events do.
Robbe-Grillet tells us that, in the Balzacian novel,

> time played a role, and the chief one: it completed man, it
> was the agent and measurement of his fate ... Passions,
> like events, could be envisaged only in a temporal develop-
> ment....
>
> But in the modern narrative, time seems to be cut off
> from its temporality. It no longer passes.... [It is a] present
> which constantly invents itself ... without ever accumulat-
> ing in order to constitute a past—hence a "story," a
> "history" ...—all this can only invite the reader ... to
> another mode of participation than the one to which he was
> accustomed.
>
> ... What [the author] asks of him is no longer to receive
> ready-made a world completed ... but ... to participate in
> a creation, to invent ... the work—and the world.[7]

To understand such plays, films, or novels, we must train
ourselves—following the visionary precepts of Rimbaud—not to ask
questions to which there are no answers. Only in this way can we
induce our imaginative powers to form a joint vision of the fragments
before us.

It is a commonplace that the artist works by his imagination; but
ever since the problem of mimesis arose, more than two thousand
years ago, the idea that art simulates something tangible has been

lurking in men's minds. This has never in fact been the case, but only modern art has made it clear that what art does is to create facts of our imagination. It is these facts of our imagination, presented to us by our artists, that form part of the thought that makes up our culture.

But the reader may recall at this point what we said about the way a poem is detached from the author's and reader's private life by the incompatibility between its form and its prose content—an incompatibility requiring their integration into a fact of the imagination wholly incommensurable with ordinary life; we said that the same mechanism works for a painting when its perspective depth is fused with the fact of its flatness. If this is the cause of the detachment in a work of art, we must then ask how visionary poetry or visionary paintings can achieve detachment, since in them this incompatibility between form and content is dissolved by the loosening of the discipline of rhyme and rhythm, in the one case, and the destruction of a meaningful prose content, in the other. The answer lies, of course, in the very fact that a visionary poem makes no coherent assertions and hence cannot become involved in expressing the purely personal interests of the poet—or the reader. The situation is similar for surrealist painting, as it is for all visionary works of art: the sort of integration they require places them directly within a field of the imagination wholly detached from the personal concerns of both painter and public.

The new poetry and art, consisting of a visionary sequence of incompatible fragments, lends itself very easily to an attack upon the incoherences of our social existence. This it did at the very start in *Fleurs du mal*. This was the time when the bohemian outlook was formed and both Baudelaire and Rimbaud contributed to it. The bohemians professed contempt for current moral principles on the grounds that the acceptance of any values not created by oneself (i.e., any already subscribed to by one's society) was intellectually dishonest, and particularly so since the society that taught these virtues did not itself live up to them. All manner of conformity was assumed to be rotten. By the end of the century, bohemianism, deepened by the reflections of Nietzsche, had developed into philosophic nihilism; and, as we have seen, the terrible disorientation that followed the First World War turned the bohemians into revolutionaries, the "armed bohemians," who threw themselves into the battles of the twentieth century which destroyed Europe.

Because painters and poets condemned the world as absurd, they represented it as a heap of fragments. But because they were artists, their vision brought this supposedly dead pile to life in their works of art! These artists thus preserved the honor of their nihilistic protest by cutting the world to pieces; but they inadvertently triumphed over this destruction of meaning in our social life by evoking in this rubbish meaningful images never witnessed before. This triumph at once crowned the artists as creators of meaningful visions and succeeded in allowing them, in their own minds, to leave the "pile" there as an expression of protest against the chaotic conditions of the age.

This triumph can be understood as the aesthetic face of our current moral inversion. Modern art may be accused therefore of having contributed to the very destruction of coherence which formed the grounds for its discovery of novel realms of the imagination. It may indeed have had some share in this destruction of coherence. But modern nihilism and moral inversion had emerged and achieved influence some time before Baudelaire first responded to them. Their basic causes lay, as we have seen, elsewhere than in the arts. The materialistic interpretation of moral principles in the *Communist Manifesto* is an indication of their development in the West, and Turgenev's *Fathers and Sons* shows them to be the ruling ideas of the Russian intelligentsia in the East, at the time when their expression in poetry had hardly begun and their influence on painting and drama still lay in the distant future. The proximate causes of our distrust of noble sentiments lie, as we have seen, in the destructive aspects of the modern world view concocted by our scientific imagination. Romanticism played an important part in exalting the absolute rights of the individual; but as the nineteenth century advanced, the romantic influence lost its battle against scientism and was swallowed up by it. The Freudian interpretation of human thought shows us how romantic individualism changed into a biological dynamics in which the "individual" of Romanticism was smashed to bits. The effects of such a disintegration of the person upon modern thought, including art, has been tremendous.

Modern art has clearly been influential in discrediting all affirmations of noble sentiments, and we may regret this; but this baleful influence does not efface its achievements. It accentuated the decomposition of meaning by crying out against it, but its power to transcend this decomposition by new ranges of visionary experience has revealed to us new worlds of the imagination. On balance, therefore, it would seem to have achieved more meaning, in spite of itself, than it has destroyed.

The structure of art will become more manifest if we look at the way similar techniques are at work in other kinds of human thought. Among these we shall take as our first example the celebration of festivities. Celebrations are a kind of action that expresses festive ideas, so a few words must be said here on the sort of imaginative meaning achieved by actions. We have touched in passing on the meaning of an actor's performance on the stage. A play does not speak of real persons or of actual events: the actor's part is to represent persons and actions imagined by the writer. He must respond by his own imagination to that of the author, and, thus guided, he must embody the part assigned to him.

We saw before that a matter of intrinsic interest may be symbolized by an object that lacks intrinsic interest. In spite of this, of course, a symbol may bear some resemblance to that which it symbolizes. Let us apply these features to an action. An action without interest of its own may acquire interest by embodying some other action of essential importance and become, therefore, a symbolic action. The meaning of a symbolic action is enhanced, however, if the symbolic action also bears some resemblance to that which it embodies—to that which it stands for. Such a symbolic action then becomes metaphorical as well. This is rather obviously the case when one makes the sign of the Cross or pours water on someone in the Christian rite of baptism. If one pays attention only to *certain* motions of the hand, one sees in the motion a figure resembling the outline of a cross; the pouring-on of water resembles the ordinary utilitarian action of washing.

Feasts and pageants and rites of mourning are not only symbolical, therefore, but are also metaphorical actions, and so their structure is similar in this respect to that of poetry and other works of art. All such occasions, like poetry and other works of art, break into the course of our current occupations and set free the imagination from the cares of the day. In painting and drama the basic techniques and instrumental material used by these arts detach them quite definitely from the course of our normal experiences. One does not, as we have noted, mistake paintings and plays for parts of our normal practical experiences. For poetry this insulation is accomplished by the poetic structure of the text. The mechanisms that serve to arouse us from our private concerns and to open our minds to follow a work of art are artificial products: their power to arouse and isolate our minds lies in their artificiality, which sharply clashes with our day-to-day experience. This is also true for feasts and solemn occasions; it is their artificial character that breaks into our daily lives and arouses our minds to other thoughts. This interjection into our daily lives is more direct here than it is for the arts; it simply decrees a pause in our regular pursuits and demands that we put on our best clothes (or some other customary attire) and take part in appropriate rituals.

Solemnities differ from works of art, however, in being essentially unoriginal. The themes of commemorations are conventional, and their forms are traditional; but the themes are important, and the forms are often deeply moving. Helmut Kuhn writes: "What we celebrate in a feast and consecrate in it can be of many different kinds, but it is basically always the same: it is the content of truth in our existence or in the existence of a society." And he discerns profundities even in the light-hearted celebration of an ordinary birthday: "When we celebrate and solemnize the passage of our life, we confirm thereby the whole natural order, of which human life with its cycle of birth and death forms part."[8]   Subjects that lie deepest in our existence are most fitly recalled in traditionally recurrent forms, since an "established" way of doing so expresses our affiliation to a comprehensive and lasting framework much better than a form we simply improvise for the occasion.

Although the spirit of medieval art may have resembled closely the ways in which solemn occasions are commemorated today, artists have usually, since the Periclean Age, tried to create works of the imagination in more uniquely personal ways. Indeed, in our own century the notion of art has become closely associated with that of innovation. Even so, rituals of commemoration can be seen to be essentially akin to works of art. Let us in an analogical way follow Helmut Kuhn's thoughts on this matter.[9] Interpreting what he says in our own way, we see that, by their detachment, works of art stand outside time and hence speak to us of one single moment. Centuries may pass over a work of art; but if it be still recognized, it will still speak of the same moment. Our feasts, ceremonials, and solemnities are also withdrawn from the day and from our passing lives. Although they do speak often of the past, they are nevertheless themselves outside time in what they seek to convey. They are timeless moments, whether what they celebrate is personal or historic.

We have a special difficulty in our day in truly dwelling in formal rituals and customs. Our modern temper balks at all such things. Since they are essentially unoriginal, we tend to deem them incapable of expressing genuine feelings and to reject them as shallow pretense. But this, of course, totally misses their point. It is the very artificiality of traditional forms that enables them to act as a framework, detaching the events to which they apply and thus endowing them with a forceful and lasting quality through the work of our own luminous imaginative powers. The destruction of formal occasion in the name of authenticity has the effect of diffusing our existence into scattered details, deprived of memorable meaning. Only through our surrender to such occasions do we find ourselves affiliated to a comprehensive, lasting framework which gives meaning to our life and death and to the myriads of separable events in between. Otherwise we do not see the universality that we share with others. Occasions, such as our birth and death, and those of others whose lives we share, can be seen as essential to a lasting whole of things when marked by appropriate ceremonies or rites. But without such ceremonies they become no more significant than the stone we stumbled over in the path or the coin we lost in the subway. Each of the numberless events in our lives is then adventitious, and the whole is inchoate and merely a tale told by an idiot, full of sound and fury, signifying nothing.

But the basis for our meaningful participation in rites and ceremonies lies in myths that in some sense we believe to be true. And here also there is a great difficulty for the modern mind. It is not that we do not have myths, but rather that those reductionistic, scientistic myths which we do have tend, because of their nature, to destroy the meaning of *all* rites and ceremonies.

We must turn our attention next to the nature of myths other than these—and to the possibility of their truth.

THE STRUCTURE OF MYTH

CELEBRATIONS AND SOLEMNITIES INCLUDE, AS WE HAVE SEEN, certain symbolic actions that are also metaphors. Looking at myths, we see that they are structured throughout by such actions. We shall focus our attention here on archaic myths—myths that are not yet complicated by literary elaboration—in order to see more clearly how myths, as such, function in the meaning-life of man. The early stages in the history of man show us more purely the particular sort of fundamental expansion of the mind that is attained through myth.

Let us begin with the story of Clever Hans. This horse was trained to "answer" simple questions of arithmetic. Numbers on a blackboard were successively pointed to, and the horse gave correct answers by stamping its foot until the right number was reached. Critical observers could find no fault in the horse's performance until they asked a question to which they themselves did not know the reply. The horse then went on stamping away senselessly. This led to the discovery that the horse had been able to stop at the right point previously because observers had unwittingly, by small gestures, signaled to the horse their expectation that he *would* stop at that point. Since he was rewarded when he stopped at these signals, the animal learned to stop at these signals. This shows how well animals can rival and even surpass man's native capacity to establish regularities inductively. Animals can also identify members of a species, that is, establish what functions as a conception of the species. Their restlessness in sleep indicates that they can dream. This shows that they also possess imagination.

In the case of man, evolution has added to these gifts the surpassing powers of language. Language makes thinking possible; i.e., we are enabled by language to "look before and after." Even so, some men may still remain hedged in by their surroundings, as animals are, and their thoughts may amount only to greater cleverness in mastering these surroundings. Later we shall come to the question whether archaic ideas about the nature of things are the result of different, and presumably lower, principles of thinking. First let us inquire into the nature of the intellectual achievement of the archaic mind in creating a myth.

Each animal forms the center of the interactions which define its surroundings, and every species has its own distinctive circle of surroundings. In his classic account of man's upright posture, E. W. Strauss writes that

> In upright posture, the immediate contact with things is loosened. . . . The horizon is widened, removed; the distance becomes momentous, of great import.

> The direction upward, against gravity, inscribes into space world-regions to which we attach values, such as those expressed by high and low, rise and decline, climbing and falling, superior and inferior, elevated and downcast.[1]

These upward- and outward-reaching imaginative capacities are only some of the conditions necessary for creating myths. The conception of the world as a whole goes beyond this mere reaching for further observable objects. It formulates a daring speculation, transcending *all* observable objects and extending the imagination far beyond any possible experienced horizon.

Some observers of animal behavior have noted certain strange behavior in animals that would appear to be of the sort we would call "superstitious" in man because this behavior seems to be acquired, not instinctual, and it does not seem to be relevant to the animal's successful adaptation; although it has no natural or causal relation to what the animal seems to be attempting to accomplish, the behavior is always performed anyhow, along with behavior that is naturalistically efficacious. Such errors are scattered and adventitious in lower animals, but they appear to be widespread and organized among men. They abound in archaic thought. The rise of man appears to have been accompanied by a burst of imaginative powers which made him liable to a whole *system* of errors of which animals are free. Any major merit, or even any mere social function, that we may attribute to the creation of myths must also acknowledge the acquired capacity for this particular kind of error; but let us for the moment set aside the errors entailed by superstition, for they inevitably raise the larger question that we are postponing, namely, the question whether the intellectual mechanism of archaic man was basically different from that of modern man.

Let us begin our inquiry into the nature of myth as an attainment of the human mind by drawing a parallel between the structure of myths and that of poetry, when both are understood as agencies for evoking the imagination. Our inspection of the content of archaic myths will prepare the ground for a comparison between mythical beliefs and the world view of modern science, to be treated later.

Our guide for the description of archaic myths, and partly also for their religious interpretation, will be the work of Mircea Eliade. For corroboration from earlier sources we shall rely on Ernst Cassirer's *Philosophy of Symbolic Forms* (but not on his *Language and Myth*) and on Lucien Lévy-Bruhl's *How Natives Think*.

Eliade sharply distinguishes myth-making from other activities peculiar to the archaic mind. We can accept this division in spite of the continuity between myths and magic and the whole system of archaic errors, for continuity does not preclude fundamental distinctions. Even within the wider area of myths, Eliade selects as its distinctive part the myth of creation, and we shall follow him also in this restriction. For this kind of myth embraces a cosmic view that may recede and sometimes disappear entirely in other myths, such as those about superhuman heroes. Let us quote here an introductory passage from Eliade's *Myth and Reality: World Perspectives*:

> Myth narrates a sacred history; it relates an event that took place in primordial Time, the fabled time of the "beginnings." In other words, myth tells how, through the deeds of Supernatural Beings, a reality came into existence, be it the whole of reality, the Cosmos, or only a fragment of reality—an island, a species of plant, a particular kind of human behavior, an institution. Myth, then, is always an account of a "creation"; it relates how something was produced, began to *be*.[2]

Myths of creation apply to all important, valuable matters of our life, giving exemplary models for diet or marriage, work or education, art or wisdom, all of which (in the myth) were perfect at the beginning:

> Myths . . . narrate not only the origin of the World, of animals, of plants, and of man, but also all the primordial events in consequence of which man became what he is today—mortal, sexed, organized in a society, obliged to work in order to live, and working in accordance with certain rules. If the World *exists*, if man *exists*, it is because Supernatural Beings exercised creative powers in the "beginning." But after the cosmogony and the creation of man other events occurred, and man *as he is today* is the direct result of those mythical events, *he is constituted by those events.* He is mortal because something happened *in illo tempore.*[3]

These acts of creation were performed during a time that is different from that in which our lives go by. This is the "once upon a time," the "sacred time" of mythical events.

> As is generally admitted today, a myth is an account of events which took place *in principio*, that is, "in the beginning," in a primordial and non-temporal instant, a moment of *sacred time.* This mythic or sacred time is qualitatively different from profane time, from the continuous and irreversible time of our everyday, de-sacralised existence. In narrating a myth, one re-actualises, in some sort, the sacred time in which the events narrated took place.[4]

This sacred time is renewed in rituals:

> . . . ritual abolishes profane, chronological Time and recovers the sacred Time of myth. Man becomes contemporary with the exploits that the Gods performed *in illo tempore.*[5]

It is important to note that what is involved is not a commemoration of mythical events but a *reiteration* of them—a doing-again of what was done "once upon a time." The protagonists of the myth are made present. One becomes their contemporary. This also implies that one is no longer living in chronological time but in primordial time, the time when the event first took place:[6]

> He who recites or performs the origin myth is thereby steeped in the sacred atmosphere in which these miraculous events took place. The mythical time of origins is a "strong" time because it was transfigured by the active, creative presence of

the Supernatural Beings. By reciting the myths one recon-
stitutes that fabulous time and hence in some sort becomes
"contemporary" with the events described.[7]

This distinction is confirmed by Cassirer:

The idea of mana, like the negatively corresponding idea of
taboo, represents a sphere distinct from and opposed to
daily life, of customary processes . . . They . . . represent the
characteristic accent which the magical-mythical consciousness places on objects [any objects].

This accent divides the whole of reality and action into a
mythically significant and mythically irrelevant sphere, into
what arouses mythical interest and what leaves it relatively
indifferent . . . ; the sacred does not simply *repel* it [the pro-
fane], but progressively *permeates* it.[8]

The recital of a myth is an experience that is detached from the
day-to day concerns of the reciting person in the same way as the frame
aspect of a work of art detaches us from the concerns of the day. It
raises us to a timeless moment. What happens when we accept a myth
is what happens when we listen to great poetry or a great play or view a
great painting: we are overcome by it and carried away into its own
sphere, away from the sphere in which we lived a moment ago and to
which we shall presently return. It is the kind of detachment that we
experience by observing a festive occasion or a day of mourning. The
detachment associated with rituals prescribed by archaic myths is
clearly akin to religious devotion, but we shall deal only briefly with
this relation at this point.

Archaic myth is detached from the everyday world in the way a strict
monotheism is—for example, that of the Old Testament. Our
personal involvement in the world is *with some parts of the world*,
while the conception of creation encompasses the *whole* world—the
world that lies beyond or under or through all its parts. The one is
concerned with things as parts, while the other ignores these matters
and has the totality of all conceivable experiences as its object.
Creation is the event by which all conceivable things are believed to
have come into existence; and the creator, or creators, are supernatural
in the sense that they transcend all particular matters. In this sense,
therefore, myths of creation are untranslatable into terms that apply to

things within the world. Archaic myths and the invocations of archaic myths are therefore of an intrinsically detached nature. They are wholly other than actual human experience.

We have already seen how thoughts so detached from our normal experience can deeply affect us in works of art and in the celebration of solemn occasions. But for religious thought we have to enlarge these terms. The integration of incompatibles accomplished for us by the creative powers of the imagination are as evident in religious thought as they were in the arts. For the idea of agencies existing outside the world and before its existence, but nevertheless operative on and therefore *in* the world, combines patent incompatibles; if it is conceivable at all, it is conceivable only by a feat of the imagination— as was the case also for the combination of depth and flatness in a painting, murders and nonmurders on the stage, etc. The creation of hitherto inconceivable conceptions by the combination of hitherto incompatible features is not, of course, restricted to art, poetry, and religious myth. It is a commonplace in mathematics and modern physics as well. But the imaginary entities created by means of the integration of incompatibles in art and myth go beyond those imaginary entities created from incompatibles in mathematics and physics. The latter are acceptable as *natural* integrations; the former, by contrast, must be called *trans*natural. Integrations of the sort one finds in mathematics and physics, though the incompatibles involved actually remain incompatible—in the sense that we never bring their elements into the perfect logical identity in which we can see that *A* is *B* because *B* is logically identical to *A*—come nevertheless to seem to be naturally compatible. At the time of their discovery, these integrations required considerable and obvious imaginative power to bring their parts together; but once we have become accustomed to working with them in the ordinary day-to-day world of our practical concerns, they seem quite "natural" to us. They "work" in our mundane world. The integrations of art, poetry, and myth, however, do not enter in a practical way into our ordinary lives. They do not "work" in such a sphere. They are, as we have said, detached from our daily lives. And their incompatibles *remain* incompatible. They must be joined together by a new act of our imagination every time we contemplate them. They thus appear to us to be meaningful and coherent but nevertheless to have meanings that go quite beyond the "natural."

Religious conceptions like the myth of creation are, however, different in significant ways from the transnatural achievements of poetry and art. The way these religious conceptions speak of the entire universe and of our destiny as human beings within these boundless perspectives makes them mystical by contrast with the concepts of poetry and art; it also makes them sacred, as we shall see.

Let us compare the way our mythical imagination detaches our minds from humdrum concerns with the way a work of art produces this effect. Eliade writes: "one 'lives' the myth, in the sense that one is seized by the sacred, exalting power of the events recollected."[9] In this a myth resembles a great work of art. But it differs from it in two other, interconnected points. First, a myth speaks of events *recollected* instead of events *represented*—because the events of creation are believed to be true; second, the rapture of a myth's being sacred is deemed to surpass the rapture of art. Eliade and Cassirer before him both speak of the contrast between myth and ordinary life as a contrast between the sacred and the profane.

We must ask how this is achieved. To Eliade the archaic myth is sacred because, being believed to be true, it reveals reality as distinct from the nonreal:

> . . . in reciting or listening to a myth, one resumes contact with the sacred and reality, and in so doing one transcends the profane condition, the "historical situation." In other words, one goes beyond the temporal condition and the dull self-sufficiency which is the lot of every human being simply because every human being is ignorant—in the sense that he is identifying himself, and Reality, with his own particular situation.[10]

Similar powers are attributed by Eliade to images and symbols:

> The symbol reveals certain aspects of reality . . . which defy any other means of knowledge. Images, symbols and myths are not irresponsible creations of the psyche; they . . . fulfill a function, that of bringing to light the most hidden modalities of being.[11]

But this still leaves us facing the question: Is the myth effective only metaphorically, in the manner of a poetic image? Think of the seventy thousand lines of poetry which educated English people may use to express states of mind they might otherwise find it hard to express. They recognize in their own experience the truth of the poet's lines. Is the truth of myths the same kind of truth? Do the two differ only in the *content* of the images that are evoking the truth in us? Is the image of creation merely of a kind that cannot be translated into tangible experience?

It will help us to answer these questions if we first survey the way myth is in fact "lived" by archaic people. We must look at the relation of these practices to magic and to the whole range of other typically archaic beliefs and then compare the whole archaic system with our own views, which are derived largely from modern science.

Let us first look at some examples of how myths of creation are relived. Histories of creation are recited in order to connect any new beginnings to it. Dancers—men and women—recite the chant of creation and the parents' genealogy during the birth of a prince. A similar ceremony is performed during a funeral service, which is thought to transfer the soul of the deceased into the other world.

> When a child is born among the Osages, "a man who had talked with the gods" is summoned. When he reaches the new mother's house he recites the history of the creation of the Universe and of the terrestrial animals to the newborn infant. Not until this has been done is the baby given the breast. Later, when it wants to drink water, the same man—or sometimes another—is called in again. Once again he recites the Creation, ending with the origin of Water. When the child is old enough to take solid food the man "who had talked with the gods" comes once more and again recites the Creation, this time also relating the origin of grains and other foods.

> To do something well, to work, construct, create, structure, give form, inform—all this comes down to bringing something into existence, giving it "life," and in the last analysis, making it like the pre-eminently harmonious organism, the Cosmos.[12]

The powers of the cosmogonic age are evoked in immemorial phrases and rituals which are believed to go back to those beginnings. When the symbolic actions are performed or directed by experts in ancient lore, the recalling of myths is converted into the practice of magic. In modern times the invocation of the divine presence by religious ritual performs a comparable function.

For Eliade the prime value of archaic myth lies in showing the world to be full of great meaning. Myth is an all-encompassing work of art, which, like any other great work of art, fills its subject with inexhaustible significance. In such a world man does not feel shut up in his own mode of existence:

> If the World speaks to him through its heavenly bodies, its
> plants and animals, its rivers and rocks, its seasons and nights,
> man answers it by his dreams and his imaginative life, by his
> Ancestors or his totems (at once "Nature," supernatural,
> and human beings), by his ability to die and return to life
> ritually in initiation ceremonies (like the Moon and vegeta-
> tion), by his power to incarnate a spirit by putting on a mask,
> and so on.[13]

This is essentially what Shelley called the wonder of our being, of which he said that poetry reveals it by purging our usual chaotic experience of the film of familiarity.

In this respect, myth does not seem to go materially beyond the scope of great works of art. But the connection with Shelley's interpretation of poetry recalls whole areas of mental detachment which range from sheer ecstasy to theological doctrines and aesthetic programs. Pure contemplation as practiced by Japanese Zen Buddhism aims at sloughing off our pragmatic observation of things and seeing them instead as fused into a comprehensive experience. We cease to look at objects severally and become immersed in them. They lose their usual meaning and are merged with the unfathomable intuition of the universe:

> Intuition is the faculty of comprehending the object of the
> soul in its interior relation with the cosmic totality lying con-
> cealed under the variety of the world. The task of intuition is
> to orient the inwardness of the human soul towards the
> NOTHING.[14]

Christian thought took a parallel line in the mystic theology of Pseudo-Dionysius. But the Christian mystic does not aim at NOTHING. He too seeks a visionary sight lying beyond the intelligent analysis of his surroundings, but by this *via negativa* he seeks the presence of God. Through a series of detachments, he strives for the absolute ignorance of particulars which grants union with him who is beyond all being and all knowledge. In a perfect love of God the world is revealed as a divine miracle.

The radical antiintellectualism of the *via negativa* is an attempt to break out of our normal intellectual framework and become "like little children." It is akin to the reliance on the "foolishness of God," that shortcut to the understanding of Christianity, of which Saint Augustine said enviously that it was free to the simple-minded but closed to the learned.

I have refrained from mentioning the Yoga, which antedates both Zen Buddhism and mystic theology, because its contemplative exercises seek complete extinction of the senses. Nirvana is destined to release us not only from the intellectual framework of perception but from our very existence as individual transmigrating beings. It is, however, in early Indian thought that we find the first theories of the union of opposites as the ultimate foundation of the world. Opposites may conflict, but on a deeper level they are one: "On the one hand there is a distinction . . . and conflict between the Devas and the Asuras, the gods and the 'demons,' the powers of Light and of Darkness. . . . But, on the other hand, numerous myths bring out the consubstantiality or brotherhood of the Devas and Asuras."[15]

In the West, early attempts to discover a unified reality underlying the manifold appearances of the world were made first by the Ionian philosophers. Later, the predominant Christian theology imposed the mystic unification of Manichaean and Arian dualities. According to Eliade the union of incompatibles was first elevated to a general theological principle by Nicholas of Cusa under the influence of the *via negativa* of Pseudo-Dionysius. He called it the *coincidentia oppositorum* and argued that such a *coincidentia oppositorum* was the least imperfect definition of God.[16]

Returning to the cult of rapturous contemplation in Zen Buddhism, we meet with a theoretical development of it into a doctrine of aesthetics. Art, poetry, and painting are said to be the transmission of visionary experience and hence to tell of the NOTHING, the Absolute, the ultimate reality. A culture informed by such visionary powers can see beauty in every gesture and every look, however humble. It cultivates an omnipresent graciousness.

Of all ancient systems of ecstatic contemplation, Zen Buddhism alone applies directly to the creative arts. Since it analyzes both the making and the appreciation of art in visionary terms, it appealed to schools of modern art as art became increasingly visionary. Moreover, modern schools, refusing the beautification of useful objects by added decorations, found support in Zen for the view that the beauty of practical objects must include their efficiency.

We have so far attended only in passing to the question of how these facts of the imagination become so fully detached from our daily concerns. That a myth of creation deals with the world as a whole, the idea of which transcends any conceptions of parts of the world, does isolate myths to some extent, as we have mentioned; moreover, the ritual observed in evoking mythical stories provides further detachment from the common concerns of the day. But when we turn to the practice of pure contemplation, which passes from the normal viewing of a landscape to a mystical contemplation of it, we do not seem to be crossing any conceptual barrier or setting up any artificial framework to separate this experience from the way we ordinarily view scenery. Where then, in such contemplation, do we find the source of detachment?

An answer to this question may be found most easily in Zen Buddhism, and this should throw light on the whole range of other mystical visions. Zen is acquired by prolonged, arduous training:

> In order to experience [Zen] we must enter a Zen monastery and take part in ... [Zen meditation exercises] under the guidance of a profound and experienced master. We must learn through control of breathing to attain unity of soul and body, and at the same time to feel incessant shocks within.[17]

Descriptions abound of the harsh discipline to which the Zen novice submits. His enlightenment is associated with the effort and the suffering of this discipline, which detaches his life from the flow of normal experience and opens to him access to ecstatic meditation far removed from the humdrum interests of life.

However grim the training of Zen may be, it is incomparably less painful than the process of initiating a youngster into the knowledge of archaic myths. There is no purpose here in describing such ceremonies; they are well known as *rites de passage*, marking the youth's entry to manhood. But Eliade tells us that they involve more than passage from one age group to another:

> The initiation goes on for years, and the revelations are of several orders. There is, to begin with, the first and most terrible revelation, that of the sacred as the *tremendum*. The adolescent begins by being terrorized by a supernatural

reality of which he experiences, for the first time, the power, the autonomy, the incommensurability; and, following upon this encounter with the divine terror, the neophyte dies: he dies to childhood—that is, to ignorance and irresponsibility. That is why his family lament and weep for him: when he comes back from the forest he will be another; he will no longer be the child he was . . . ; he will have undergone a series of initiatory ordeals which compel him to confront fear, suffering and torture, but which compel him above all to assume a new mode of being, that which is proper to an adult— namely, that which is conditioned by the almost simultaneous revelation of the sacred, of death and of sexuality.[18]

Even allowing for the possibility that Eliade's points may be somewhat overdrawn, it is clear that initiation ceremonies effectively stamp in the minds of these people the esoteric nature of archaic myths. This should suffice by itself to establish the detachment of their imaginative vision.

But the overpowering force of initiation ceremonies brings back and reinforces our doubts whether myths of creation can be said to be true and their substance real, as Eliade claims. Can that be real and truly meaningful which requires such terrible methods of indoctrination? We must deal with this question of truth in myths in our next chapter.

9

TRUTH IN MYTHS

IT IS AN OBVIOUS FACT THAT ARCHAIC MYTHS FORM PART OF A WIDE system of archaic beliefs that bristle with absurdities. In his classic study, *How Natives Think*, Lévy-Bruhl argues that the representation of the world accepted by primitive peoples differs basically from the views that science has taught to modern men.[1] He thinks that the cruel shock of initiation ceremonies is bound to suffuse the teaching transmitted to the adolescent with emotions that obscure his mind. The very opposite of this view has apparently been put forward by Lévi-Strauss in *The Savage Mind*, where he holds that the mechanism of reasoning is inherent in the physical structure of matter and hence must be the same in the minds of both primitive people and modern men.[2] This is closer in principle to the position of the earlier English anthropologists, particularly Tyler, Frazer, and Andrew Lang, whose views Lévy-Bruhl sharply rejects.

The difficulty with Lévi-Strauss's view is that the mechanical process of thought deemed by him to be universal is only a crude approximation to the actual process of acquiring knowledge, which is based ultimately, as we have seen, on tacit integrations that have no machine-like equivalent. Lévy-Bruhl appears to be right in holding that the primitive mind acquires its strange views directly from its unique perceptions and not by deriving faulty conclusions from "correct" perceptions, i.e., from the kind of perceptions that modern men would accept. We must nevertheless dissent from him. The cognitive process in both the primitive and the modern mind, we think, *is* rooted in the same principles, but the results differ because archaic thought tends to be based on more far-reaching tacit integrations than are acceptable to the scientific mind of modern man. We may agree that many of these primitive integrations are nonsensical; but *some* sort of tacit integrations are, as we have indicated above, truly essential as a basis for all knowledge. Therefore, a scientific method aiming at dispensing with tacit integrations altogether is also nonsensical. It is on these grounds that we shall be able at last to answer the question in what sense, if any, we can accept Eliade's view that archaic myth is true.

We shall approach the question concerning the truth of myths tangentially in this chapter by first examining archaic beliefs which deal with rather ordinary day-to-day events but do so in ways quite different from those in use in the modern world. These beliefs are not unconnected with myths, but they seem to have a more secular content than the myths themselves. They can thus be understood to have something like a functional equivalency to our own ideas about matters of daily experience—ideas which are based on modern science and technology. Our recognition of the similarities present within the differences should then provide us with a key to the nature of our problem about the truth of myths.

Let us pursue the idea that the absurdities of archaic thought are due to an excessive use of the same integrative powers that modern thought applies (on the whole more wisely). First let us recall the many uses to which such integrative powers are put. We have seen that they are used in perceiving, in inducing muscular coordination, in practicing skills, in establishing semantic relations, in forming conceptions, and in creating works of art. We have also seen that they underlie the processes of scientific discovery and of technical invention. Some of our integrative actions require little effort on our part; merely looking at an object usually suffices to induce the integration of its impressions on us. Other integrations demand persistent efforts by exceptionally gifted minds. All integration must rely on the services of the imagination, but the imaginative powers at work become substantial, and indeed massive, as the elements to be integrated become increasingly disparate.

We have spoken at some length of the integration of elements which, when focally observed, appear incompatible. Such integration may be almost wholly spontaneous (as in perception) and require little imagination (particularly in binocular vision). On the other hand, it may require special gifts (as in the creation of visionary art) or rely on prolonged training (as it does, for example, in achieving the visionary experiences of Zen Buddhism).

This brief recapitulation should suffice to demonstrate that whether integration results in establishing perceptual or cognitive facts or in producing works of the imagination, the process is essentially informal. The fact that a coherence established by integration will have qualities *not present in* the subsidiaries used in composing this focal result is in itself proof of this, since in a *formal* process the result can be seen to have been fully present in its antecedent premises, i.e., to be a logical implication of them.

It follows that there can be no strict—i.e., formal—rules for accepting or rejecting the validity of an integration. In many cases, of course, there is practically no choice in this respect. To identify an animal as an Indian elephant does not involve much choice. But integrative inferences of great importance may appear plainly acceptable to some and just as plainly absurd to others. Take the case of Dr. Immanuel Velikovsky. In his *Worlds in Collision* Velikovsky put forward a theory, based on acceptance of reports from the Old Testament, the Hindu Vedas, and Greco-Roman mythology, of catastrophic events in the earth's history from the fifteenth to the seventh century B.C.[3] To account for these events, Velikovsky supplements the force of Newtonian gravitation with powerful electrical and magnetic fields acting between planets. His book was widely acclaimed and became a best-seller, even though it was angrily rejected by scientists. A number of sociologists supported the popular view against the scientists. They came out first in the *American Behavioral Scientist* and then again in a book by Alfred de Grazia, angrily attacking the whole community of natural scientists for paying no attention to Velikovsky.[4] For my part, I believe that the scientists were quite right in refusing to pay serious attention to Velikovsky and that the sociologists' attack on them was totally unfounded.

I argued this in a paper in *Minerva*, republished in *Criteria for Scientific Development*.[5] I said that "a vital judgment practiced by science is the assessment of *plausibility* . . . which is based on a broad exercise of intuition guided by many subtle indications." The difference between those who take Velikovsky's theories seriously and those who reject them out of hand lies in their very different assessments of the plausibility of these theories. And since plausibility cannot be formally demonstrated, the conflict between the two sides has remained unresolved for more than twenty years.

Though more far-reaching, most differences between modern thought and that of archaic people are of the same kind as the differences between modern astronomers and the followers of Dr. Velikovsky; for in both cases the differences have to do with judgments of plausibility.

Many of the strange observations made by archaic people are clearly based on the same kinds of inference we use today. Archaic people observe causal relations, and many of their observations of this kind correspond exactly to our own. They are based, like our own observations of causality, on a temporal sequence of events or a recurrent contiguity of facts. But archaic people may draw fantastic conclusions from these contiguities by being much less prepared than we are to regard them as merely coincidental. However, we must remember that, for thousands of years, predictions made by astrologers from the position of the sun, moon, and planets in the constellations at a man's birth were accepted as valid, though such relations are purely coincidental according to our quite well-accepted modern scientific views of the stars.

When we are told, therefore, that the principles of *post hoc ergo propter hoc* and *juxta hoc ergo propter hoc* are characteristic of mythical thinking, we must reply that these are the proper guides of all empirical thinking, as even David Hume himself pointed out, and that modern man differs from his archaic ancestors only in judging whether certain observed temporal or spatial contiguities should be deemed coincidental or causal. It is only owing to what we do indeed believe to be our more correct view of the general nature of things—derived largely from science—that we apply the principles of causation more aptly (to our own way of thinking) than primitive people do. Even so, fears of magical causes persist even in modern intellectual circles. For example, in the Athenaeum Club, center of distinguished scholars and writers in London, the numbering of bedrooms avoids the figure 13 by calling it 12A. The fear that sleeping under a number 13 may bring misfortune has not vanished altogether. In addition, of course, the expectation of magical results from the invocation of mythic origins has its counterpart in Christian religious services, which invoke a divine presence. They are matched even more widely by the precatory prayers of all times.

However, some curious cases of magic rely on relations differing in principle from any that are considered possible today. One example is the magical control of persons and sometimes of divine powers by a knowledge of their secret names. As Cassirer puts it:

> [In] mythical thinking the name ... expresses what is inner-most and most essential in the man, and it positively "is" that innermost essence. Name and personality merge.... He who knows the true name of a god or a demon has unlimited power over the bearer. [Moreover, a] man's image, like his name, is an alter ego: what happens to the image happens to the man himself.... If an enemy's image is stuck with pins or pierced by arrows, he himself will suffer immediately. And it is not alone this passive efficacy that images possess. They may exert an active power, equivalent to that of the object itself. A wax model of an object is the same and acts as the object it represents. A man's shadow plays the same role as his image or picture. It is a real part of him and subject to injury; every injury to the shadow affects the man himself. One must not step on a man's shadow for fear of bringing sickness upon him.[6]

A name, an image, a shadow—these things are what they are because a person exists and they bear on that person. Nor is the being of the person unaffected by such things bearing upon him. If we know his name, we can call him by name, we can talk about him and think about him. A named man is enlarged by his name as an object is enlarged by adding a handle to it or as a person is transformed into a more inclusive entity by his attire. And so a person is also enlarged through the creation of a painting or a wax model that bears upon him. He even becomes something larger through the mere existence of his shadow, for it originates in him and thus points to him as a shadow-casting body. Anything that intimately bears on a person bears in the same way that the parts of a whole bear on the whole which they form.

The idea that a name, an image, and other attributes that bear on a person are real substitutes for the person has been called the principle of the *alter ego*. Closely akin to this principle, and indeed comprising it, is the principle of *pars pro toto*, the archaic mind's way of identifying a part with the whole to which it belongs. Let us refer again to Cassirer:

> The whole does not "have" parts and does not break down into them; the part *is* immediately the whole and functions as

such. This relationship, this principle of the *pars pro toto* has also been designated as a basic principle of primitive logic. However, the part does not merely represent the whole, but "really" specifies it; the relationship is not symbolic and intellectual, but real and material. The part, in mythical terms, is the same thing as the whole, because it is a real vehicle of efficacy—because everything which it incurs or does is incurred or done by the whole at the same time.

Anyone who acquires the most insignificant bodily part of a man—or even his name, his shadow, his reflection in a mirror, which for myth are also real "parts" of him—has thereby gained power over the man, has taken possession of him, has achieved magical power over him. From a purely formal point of view the whole phenomenology of magic goes back to this one basic premise, which clearly distinguishes the complex intuition of myth from the abstract, or more precisely abstracting and analytical, concept.[7]

This description can be recast and expanded by means of the structure of tacit knowing. Anything that bears on something else, whether as designating it or being part of it, is known to us, in this connection, subsidiarily. Viewed in this from-to relation, it has a meaning which lies in the focus to which we are attending from it. The meaning of a subsidiary is wiped out if we turn our direct focal attention on it, and this destruction of the from-to relation changes the appearance of both its subsidiary parts and its focal point. These changes are best known in the dismemberment of a figure which results from our attending focally to its parts. In fact, these changes form the very foundation of gestalt psychology.

These from-to interactions apply equally to all other meaningful relations. The archaic mind seems sharply impressed by the sensory qualities of meaningful relations, and its imagination greatly exaggerates the interaction between subsidiaries and their focus. We meet with no evidence of this exaggeration in the from-to relation between the names of inanimate things, like bows, arrows, or rivers, and the things they designate or between the parts of broken objects, like the shards of a broken vase and the vase itself. It is only when the focus of the semantic relation is a living being, and particularly when it is a human person, that the imagination seems to enlarge so greatly on the dynamic forces involved in the subsidiary-focal structure of integrations. Of course, the vastly indeterminate manifestations of a human being may well lend to him a mysterious quality which often stimulates the imagination far beyond the range of experience. It is most interesting to note at this point that the modern mind errs in the opposite direction. Its conception of meaning *fails* to note the deep-set qualities of from-to relations and seeks to reduce the human mind to a predictable system of responses.

We shall return to this later. Meanwhile let us enlarge further on the archaic mind. The same kind of difference between the archaic and modern views of nature is sharply evident in yet another point, namely, in the archaic mind's identification of man with animals, particularly in totemism. Here we encounter still another feature of the structure of tacit knowing. All observation, as we have seen, is based on our interpretation of the reactions of our sense organs, and also of deeper parts of our body, to the impact of external stimuli. These bodily reactions are experienced subsidiarily; what we perceive is the joint meaning of these subsidiary reactions. The moving of our limbs has a similar structure, as we noted in chapter 2; it consists in mobilizing a set of muscles that jointly bear on our deliberate movements, and this movement is in fact their joint meaning.

Our perceptions and deliberate bodily motions have therefore the same structure that we have ascribed generally to meaningful relations. But to observe a meaningful relation is to integrate its subsidiaries as bearing on their focus; hence it is to handle these subsidiaries as if they were responses inside our own body. In this sense the structure of such an observation is, as we have seen, achieved by our dwelling in its subsidiaries. Generally, all comprehensive entities are known by our dwelling in them, and to this extent we participate in them as if their subsidiaries were parts of our body. Indwelling is more accentuated in the case of our knowledge of living things, particularly human beings and animals having a structure similar to our own. For in these cases we dwell in, and relive thereby, the very motions by which the person or animal carries out its actions. This is no less than a sharing of lives between us and other men, as well as between us and the higher animals.

This sharing of lives is not due to a deliberate effort on our part. It might be most unwelcome to us. Medical students must harden themselves against being overwhelmed by the spectacle of the surgical opening of a human body. We can see here also the powerful effects of the very *pars pro toto* which we regard as so strange in archaic people.

The dead body of a person dear to us moves us deeply. We do not find Antigone's outrage absurd when she sees her brother's body thrown to the dogs. I have seen a first-year medical student faint at the sight of a thighbone freshly carved out from a cadaver. Operations on animals are almost equally disturbing.

There is strong evidence here for the view that our knowledge of other persons, and of animals too, is based on participating vividly in their lives. If we ourselves know animals and men in such a profoundly stimulating manner, it seems understandable that, at a time when much of life consisted in hunting or being hunted, men's imaginative participations in the life of animals would tend to go well beyond this and become altogether totemistic.

It admittedly seems quite absurd when members of a Brazilian tribe called Bororos declare with pride that they are a kind of red parrot,[8] and we can well understand that Lévy-Bruhl regards such identifications as evidence of a distinctive prelogical power in the archaic mind, which he calls *participation*. Though we have found it essential to use this same term to account for our knowledge of animals, as such, the sense in which we use it in this context is quite different from the sense in which it is used to account for the fact that archaic tribes identify themselves with a particular kind of animal.

However, the identity claimed by totemism does not mean that the two things, the man and the animal, are interchangeable. A Bororo never mistakes a red parrot for a Bororo tribesman. On the other hand, it is not unusual for modern man to identify two objects that he can easily distinguish from each other; he does this all the time when he identifies different members of the same species. Bororos seem to think that in some ways they and the red parrots belong to the same class. But isn't this kind of identification a manifest absurdity? Not, of course, if it is within the range of the Bororos' particular sense of plausibility. Many scientists and philosophers have, for centuries now, asserted that all human beings, including themselves, are automatically functioning machines. Some modern thinkers have made this even more telling by arguing that machines possess consciousness and can have every kind of human feeling. To many of us this identification seems absurd. To them, on the contrary, the view that men have minds which control their actions is the absurd view. What the Bororos mean by identifying themselves with red parrots may be difficult to fathom, but there is no necessary reason to say that it is any more absurd than the view of many scientists and philosophers that they are machines.

Eliade's view of archaic myth as a source of truth must be viewed in this connection. It is based on the rejection of the modern scientific outlook, which is also, as we have seen, a set of beliefs. Let us review here how these "strange" beliefs belonging to the scientific world view arose and how they became so firmly impressed on the modern mind.

All science must surely grow from prescientific thought, and therefore the articulate machinery of science must originate from largely tacit antecedents. Biology has resulted in broad confirmation of its antecedents. Physics, however, has developed by rejecting or bypassing them.

Let us look at biology first. Fundamental features of life were commonly known in a tacit and inarticulate way before biology developed. Animals and plants were recognized before the development of zoology and botany, health and sickness before the science of pathology came into existence. The contrast between sentience and insentience, between intelligence and its absence, were known before they were studied by science. These were common knowledge, and so were many details of living functions: hunger for food, the need to breathe, the processes of digestion, elimination, and secretion, the functions of our senses, the processes of procreation, of embryological development, of growth and maturation, of senescence. One could go on indefinitely enumerating the subjects which biologists took over from popular knowledge.

Large branches of biology now relate to different sections of life that were identified only tacitly before the rise of science. These scientific studies have modified and supplemented the tacit conceptions by which they were guided, but they have rarely superseded them. Their richness testifies to the perspicacity of the insight that led to their original formation.

The characteristic shapes of living things have no geometrical definition and are simply recognized as physiognomies. The same is true of the various living functions, which are physiognomies extending in both space and time. Our prescientific knowledge of living things must therefore have arisen from acts of profound indwelling, comprehending the general panorama of biotic features—*and this remains the way these features are known today even by scientists.*

Biology must ultimately bear on life as life is known to the nonbiologist. Otherwise it has lost any claim to a specific subject matter. It must contribute to the explanation of the features that form the generally known panorama of life. These features have at least one thing in common, namely, *they can go wrong*. This singles them out as *achievements*—as against atoms, molecules, and planets, for which such a term is clearly inappropriate. Attempts to understand an achievement must ask the question: How is it accomplished? What may cause it to go wrong? Biology consists predominantly of answers to such questions. It seeks to discover the mechanisms by which life works.

The same logical structure is found in the science of engineering. We may be in the habit of using a watch, an automobile, or other machinery without knowing how it works. To find out how it works, we must take the object apart, map out the relation of its parts, and discover the way each functions in keeping the mechanism going. Ancient crafts exist, like brewing and pottery, which have developed empirically. Inquiries undertaken to find out how these processes work have exactly the same logical structure as inquiries into how physiological performances work. Some of Louis Pasteur's major biological studies were indeed these kinds of technological investigations. Biology is, in effect, the technology of life.

Such biological analysis into mechanisms alternates with integrations of these into greater meaningful wholes. New comprehensive entities may be established in the manner in which Harvey discovered the circulation of the blood or Mendel discovered that heredity is atomistic or the process of evolution was established from a great range of clues bearing on it. But after the discovery of a new comprehensive entity, we always ask once more: How does it work? The science of biology can be greatly expanded by phases of integration, but it always aims beyond these at further solutions of fresh analytic problems. Each new integrative discovery results in a further "how."

The practical advance of engineering and technology consists primarily in the invention and construction of devices that will work. This too is an integrative pursuit. It skillfully combines specially devised objects and processes to form a useful mechanism. The task is conceived in the imagination and is completed by actually producing the mechanism and making it work.

Turn now from the panorama of life and of human contrivances to the spectacle of inanimate nature. But where do we find this spectacle? There are many interesting inanimate objects about us most of the time, but practically all of them are contrivances of man. Even when we look at our gardens, some soil and traces of rainwater are almost the only visible representatives of inanimate nature. We must think of some desolate region, of rocks, sands, rivers, seas, of the winds, the clouds, the sun by day and the moon and the stars by night. This was therefore about the range of conception men had of inanimate nature before science. Compare this with the wealth of intriguing items from which the study of biology originated and on whose interpretation its ultimate interest continues to depend. The only distinctive pre-scientific features of the inanimate world were the stars, the sun, and the moon, together with their curious motions. Few inanimate objects were known that puzzled men by their distinctive shapes or behaviors. Nothing suggested the hidden beauties of the gas laws, of thermo-dynamics, optics, and acoustics; of the potential presence of atomic spectra; of a thousand elementary particles and a million organic compounds; of Newtonian mechanics, of quantum theory and relativity. And so this great intellectual system of physical science does not derive its interest from the light it throws on its prescientific antecedents.

Such intellectual stimulus as inanimate nature has provided to scientific inquiry has come almost exclusively from watching the skies. This stimulus was intense and eventually brought beautiful results, crowned by Newtonian gravitation and general relativity. But before this occurred, the misleading impression that the Earth was at the center of the universe had dominated thought for many centuries. This notion, expressed in Babylonian astrology and later in Aristo-telian cosmology, imbued man's outlook with a host of erroneous suggestions. When the Copernican revolution had finished refuting these ideas and had inspired in their stead a great new system governed by natural laws, the common man's confidence that he could understand the world in terms of his ordinary experiential perceptions of it was shattered. He agreed henceforth to accept unconditionally the scientific view of things, however absurd some of it might appear to him.

Ever since Laplace first raised the point in defining Universal Knowledge, philosophers have discussed the notion that from today's topography of the ultimate particles of an object we can predict, by the laws of mechanics, any future topography of these particles. The immense difficulty of carrying out such computations is easily perceived. This has diverted attention from the far greater difficulty involved in the idea itself, namely, that the results of such a calculation would in themselves tell us nothing of any importance. Admittedly, some observable features could be derived from them. By adding up the energy of the particles, we could estimate changes of the heat contained in different areas. But the computation of temperature changes would be beyond our power, since we would not know what probability (i.e., entropy) to attribute to any particular section of the topography.

Let us assume, however, that these kinds of difficulties could somehow be overcome, so that we could derive a complete physical-chemical topography of our object. Could this provide us with a comprehensive, meaningful knowledge of all objects?

Such a topography would still be almost meaningless in itself. How far can we go beyond it? If the topography is that of a living being, say a frog, can we recognize the frog from it and derive the mechanism of the frog's vital functions?

We have seen that living beings are characterized by their physiognomies, including the space-time physiognomy of their functions. A physiognomy obeys no mathematical formula; it can be recognized only tacitly by dwelling in its numberless particulars, many of them subliminal. We can know a frog only by dwelling in its particulars in this way, and we can know the topography of a frog only if we are able first to know *a frog*. We can know the emotional and intellectual life of animals only by an even deeper indwelling, deep enough to achieve an empathy with their consciousness. And suppose that we study a great human mind: we can enter into its thoughts only by respectfully submitting to its guidance.

To attribute such levels of existence as these to an atomic topography seems as absurd as it would be to talk about the smell of differential equations, yet the modern mind seems hardly to hesitate in countenancing such incongruities. Cowed by the experience of the Copernican revolution, we dare not trust the testimony of our senses to contradict the teachings of science. To "doubt the sun doth move" was still the epitome of absurdity to Shakespeare seventy years after the death of Copernicus. About this time Galileo spoke triumphantly of the acceptance of Copernicanism as "the rape of the senses." By the age of Kant the Newtonian view of the world, as consisting entirely of hard masses in motion, had gained the status of a necessary framework of experience. Man's acceptance of a seeming absurdity had ushered in the triumph of an actual absurdity.

Our account of archaic thought confirms Lévy-Bruhl's rejection of the theories of Tyler, Frazer, and Andrew Lang. It also confirms Lévy-Bruhl's view that the primitive mind acquires its strange views directly, in the very process of perception, and not by deriving mistaken conclusions from the kind of perception that modern men would accept. Though we quoted mostly from Cassirer's more ample observations on archaic thought, Lévy-Bruhl's earlier account of it must be greatly admired. His simple comparison of it with the view current today is preferable to Cassirer's analysis, which reaches very far for an explication of it in the Kantian terms of pure perception compared with a perception disciplined by the categories of the understanding.

But we can hardly agree with Lévy-Bruhl that primordial minds used some different "prelogical" modes of inference, or at least we must put the matter differently. All empirical observation rests ultimately on the integration of subsidiaries to a focal center. All such integrations—from perception to creative discoveries—are impelled by the imagination and controlled by plausibility, which in turn depends upon our general view about the nature of things. Over a wide range of day-to-day affairs the archaic mind thinks and acts as sensibly as we do. Lévy-Bruhl himself says this clearly.[9] In certain respects the archaic view of things is of course different from what is loosely known as the modern scientific view; but while the range of difference may be wider, the difference itself is perhaps not deeper than the differences in views about the plausibility of coherences held by various groups of people in a modern Western university.

In evaluating the differences between the archaic and the modern approaches, we have to maintain that the archaic mind is better in many ways. It is right in experiencing names as part of a named person and an image as part of its subject; for a name is not a name, nor an image an image, except as a subsidiary to the focal center on which it bears. And such is the nature of all meaningful relations. Admittedly, the archaic mind tends to exaggerate this coherence to the point of absurdity, but it is closer to the truth than the modern view, which has no place for the quality and depth of these coherences nor, therefore, for the full extent of the subsidiaries that are necessary to their composition. This difference becomes essential in the observation of those comprehensive entities that can be observed only by indwelling. The archaic mind recognizes indwelling as the proper means of understanding living things. Modern biology and psychology abhor this approach to life and mind. The import of their teaching tends rather to be that we are all machines and, in the last analysis, mere atomic topographies. These ideas of Galileo, Gassendi, and John Locke, coupled by Humean associationism, have paved the way to the achievements of modern science, but at the same time they have deprived everything that is of primary interest in the world of any grounds of meaning for us.

This finally brings us back to the question to what extent archaic myths of creation can be said to be true, as Mircea Eliade claims they are.

We have shown by the example of the Velikovsky case how profoundly our judgments of plausibility affect the conclusion we draw from a particular set of data. This has also served as an example of the indemonstrability of plausibilities. Even within the same academic community contradictory judgments of plausibility can be upheld by different groups. This fact proves that these judgments, which underlie every empirical inference, rely to a decisive extent on grounds that are not specifiable. We have seen that the often fantastic aberrations of archaic beliefs are largely due to a reluctance to accept the possibility that some conceivable coherences are merely coincidental. We have also pointed out that our modern scientific education teaches us an even more absurd view about the nature of things when it affirms that all coherent systems of our experience—including our own conscious existence—can ultimately be represented by their atomic particles interacting according to atomic forces. This too is an aberration of the imagination: a fantastic extrapolation of the exact sciences.

The fact is that all empirical knowledge is rooted in subsidiaries that are to some extent unspecifiable. We may add to this as its corollary that the range of meaning covered by verbal statements is unlimited. We have seen how richly poetic meaning can serve to clarify our own experiences and to express them effectively. The myths of archaic people should be regarded in this light. They are clearly works of imagination; and their truth, like the truth of works of art, can consist only in their power to evoke in us an experience which we hold to be genuine.

But is this not to move too far away from what is usually considered as verification—particularly in science? Not altogether. In the history of science there are many instances of partly true ideas that were totally rejected by some scientists because of their erroneous content while others accepted them in spite of all their errors. Mesmer's fantastic demonstrations of mesmerism were successfully followed up by Elliotson; but since the latter's version of them still included an ample portion of absurdities, most scientists rejected his results as they had rejected Mesmer's, and it was not until Braid's reformulation of mesmerism by the conception of hypnosis, a century after Mesmer, that the truth content of his work was discerned and recognized as distinct from its fantastic errors. There is a myth abroad today that a scientific theory is instantly rejected if we come across any facts that are incompatible with the theory. But, as we have seen earlier in this work, the actual practice of scientists is often to doubt the validity of the demonstration of such incompatible facts, however inexplicable the evidence may appear to be; or else to include any facts apparently contradicting an accepted theory as anomalies of it; or, in yet other cases, to accept two mutually contradictory principles, ascribing to each its range of applicability in the hope that something will turn up to explain the conflict between them. So, upon occasions, even a scientific theory is accepted in general for the truth supposedly grasped *in* it, not because all parts of it are equally credible.

With this in mind, we can renew the question about the truth of archaic myths of creation. There is a great deal of nonsense in these myths. Much of this we may be able to isolate and disregard, but even the heart of the story contains much that is unacceptable. Can we suppose that this is one of those cases where a truth is so interwoven with error that it cannot be expressed without affirming the error as well? In other words, if we reject archaic myths on account of their manifest errors, will this purification not be outweighed by a concomitant loss of the truths contained in these myths?

But what are such truths? Eliade says that the myth of creation makes us aware of a deeper reality that we inevitably lose sight of in our personal pursuits. It sets us free from a false identification of Reality with what each of us *appears to be or to possess.* . . . The myth continually re-actualises the Great Time, and in so doing raises the listener to a super-human and suprahistorical plane; which, among other things, enables him to approach a Reality that is inaccessible at the level of profane, individual existence.[10]

Thus the myth of creation opens to its followers a certain view of the universe and makes them feel at home in it. In the light of the myth every major event of man's life evokes his descent from his ancestral cosmic origin, and his every major enterprise is undertaken as a rehearsal of the mythical act which first performed such an enterprise. The myth of creation teaches knowledge of perfection, of perfection in nature and of virtue in action. Its immemorial knowledge links those who possess this knowledge to an endless company of fathers. Mythical knowledge provides on sacred occasions the experience of thoughts beyond the range of men's individual lives. It secures to its disciples access to a detached experience of meaning, echoes of which will follow them into their daily life.

These results of accepting the myths of creation produce in us experiences that we can believe to be largely genuine and therefore largely true. A belief in the gradual emergence of man from an inanimate universe reveals to us that the dead matter of our origins was fraught with meaning far beyond all that we are presently able to see in it. To set aside an achievement as full of meaning as this—as if an emergence of this sort could happen any day by mere accident—is to block the normal sources of inquisitive thought.

Man's origin is a mystery which the myth of creation expresses in its own way. And the image of man's destiny, as derived from his mythical origins, is much nearer to our own experience of our own lives, to our experience of human greatness, to our perception of the course of our history since history began, and to our experience of the shattering forces of our utopias than is the image of the barren atomic topography to which the ideal of detached observations seeks to reduce these matters.

There is, therefore, an important truth in the archaic myth of creation that is missing from the present ideal of scientific knowledge, and in this sense we can agree with Eliade when he speaks of the creation myth as being true. For some of us this is sufficient. For others, however, it is not sufficient. In fact, it may be insufficient for the greater part of the present generation of human beings. We shall see why this is so in the next chapter.

10

ACCEPTANCE OF RELIGION

LOOKING BACK UPON WHAT WE HAVE BEEN DOING IN THIS WORK, we see that we have found it necessary to introduce certain new elements in order to expand our notion of knowledge from that which is related, directly or indirectly, to *observations* to that which consists of the sort of *acceptance* we grant to a work of art.

We have looked especially at works of representative art, because these always include some affirmations that can be expressed verbally in prose statements. Whenever we accept works of this kind as genuine works of art, we also understand them to be true, even when the prose statements of their content may all be admittedly false in themselves. We have investigated several modes of affirmation that are understood by us to be valid in this way, even without (or in spite of) any observational content. Representative arts are thus seen by us to make statements that we accept in some manner, even though we do not find them in the least compelling as *facts*.

One of the new elements we have found it essential to introduce in order to move from "observation" to "acceptance" is the fact that the (false) statements embodied in works of representative art are always fused with conditions incompatible with these statements. This circumstance requires, as we saw, repeated and vigorous efforts on the part of our imagination to fuse or integrate these incompatibles. In visionary art, which lacks a "story" altogether, the vital role that our imagination plays in our acceptance of art was brought into sharp relief. We saw that, although our imagination is necessarily at work in every kind of awareness or activity in which we engage, it is at its peak in our apprehension of works of art and in similar achievements.

Knowledge bequeathed to us by scientific discoveries, we saw, eventually becomes commonplace knowledge to us and seemingly requires no imaginative effort on our part to make use of it, although its original discovery may have required a great deal of imagination. A work of art, on the other hand, is meaningless to us unless we exercise our imagination upon it each time we experience it.

Another new element we found we needed in grasping the nature of "acceptance" was something we named the "frame" (as opposed to the prose content or "story"), e.g., the paint and canvas in a painting, the stagecraft in a play, the rhyme and meter in a poem. This element makes for the detachment enjoyed by a work of art. This frame, by setting the work off from the normal course of our experience, keeps the prose statements made in the work of art from being considered as relating or predicting any facts observable in our ordinary practical experience.

This circumstance led us to see that the ancient conception of works of art, in which they were thought to be imitating or simulating objects, is false. Paintings, plays, and poetry do not affirm anything that either is or is not the case, for their affirmations become integrated into frames incompatible with these affirmations. The resultant fusion cannot, therefore, convey any factual information. We saw that Coleridge's conclusion that we accept these works of art by an act of voluntary disbelief is therefore wrong. The *artifact*, fused to the so-called mimesis, deprives it of mimetic content.

Hence a representative work of art contains a built-in contradiction which distinguishes the character of its affirmation from any empirical statement. This contradiction separates its affirmation from the context of our life space—the context of the whole course of our existence—and causes it to be *detached* in this sense from both its author and its public and, indeed, from any natural experiences, including those of science. We found we could call its meaning *trans*natural, since its gross contradictions take a work of art out of both the context of our passing *lives* and the context of the *space* in which we live, whether as laymen or as scientists.

Festivities and solemn occasions we found, are also detached in this same sense, namely, through being external to the context of our lives and of any experiences at all. And we ranked, as also among these essentially detached affirmations, all sacred acts of ritual and all sacred stories recited within the context of a ritual that effectively isolates them from the flow of time and the surroundings of space.

There was also a third new element that we perceived to be operative in "acceptances" as opposed to "observations." Recall our example of a national flag as an instance of a symbol emotionally representing a country to those of its citizens with strong patriotic feelings. It is a striking fact that the whole range of lifelong recollections, diffusely remembered by the citizen of a nation, is condensed and infused into a piece of cloth attached to a rod and bearing a conventional pattern. The immensely extended, hardly recallable life of a person is condensed into an emotional force attached to an otherwise trivial, meaningless, object.

Repeat this description of a flag's symbolic powers with some variations and you have before you the emotional powers of poetry as we described them, following their analysis by I. A. Richards. Memories of our diffuse day-to-day life contain experiences that supply meaning and sharp emotional power to poems able to embody such experiences through the integrative efforts of our imagination. The same sorts of experiences, immersed in our diffuse memories, supply the meaning and power of a painting or a play. Works of representative art are as detached from our ordinary incoherent lives as a piece of colored cloth on a pole is detached from the lives it represents, and works of art derive their power through the same mechanism by which national flags derive theirs.

The example of the flag fails, of course, to take us far enough. Works of art have an intrinsic meaning that a flag has not. Works of art are in a sense metaphors; and although the power of a metaphor is akin to that of a symbol, such as a flag, there is a difference that gives more meaning to a metaphor than is present in a symbol. We saw that the power of the metaphor lies in our capacity to embody an object of principal interest (the "tenor") in another remotely similar (but also intrinsically interesting) object (the "vehicle"), thus giving the first object a new sharp and emotionally charged meaning.

It has occasionally been said that all poems are metaphors or that all paintings are metaphors. This anticipates vaguely what we have developed here, namely, the fact that there exists a class of meanings based on our capacity to embody one thing in another (as we do in metaphors); but this capacity also includes our ability to immerse a whole set of diffuse experiences in a more sharply circumscribed experience (e.g., in the experience of a single work of art) that bears some resemblance to the set of diffuse experiences. The more sharply circumscribed experience—which *can* be the vehicle in a metaphor but

which can also be a whole painting, a whole play, or a whole poem—
then absorbs and embodies our original diffuse experience, becoming
a strikingly revealing, emotionally charged interpretation of it. This is
what happens when we are carried away by a poem or are spellbound
by a play or when we first see the *Moses* of Michelangelo and are forced
to revise our estimate of the human race. Even a symbolic action, we
may recall, can also be a metaphor. It is a metaphor when the action
not only stands for but resembles that which it symbolizes, which is
the case in rites and ceremonies. These then can also carry us away,
since they are actions detached from our ordinary lives and are full of
meaning because of the sacred myths that lie behind them and draw
us into the Great Time.

We are now ready, working along these lines, to approach an
interpretation of religion. When we spoke of the structure of myths in
chapter 8, we noted that this sort of close connection between archaic
creation myths and primitive rites and ceremonies moved us into the
realm of religion, but we have not yet explored to any great extent the
nature of the integrations—the meanings—that are attainable in
religion. Religion, we can see, is a sprawling work of the imagination
involving rites, ceremonies, doctrines, myths, and something called
"worship." It is a form of "acceptance" much more complex,
therefore, than any of the other forms we have been attending to.

First of all, religion involves sacred myths that inform rites and
ceremonies, imbuing their intrinsically metaphoric meaning with
something more than the kind of poetic or artistic meaning they
would possess simply as metaphorical works of art. Let us take the
Christian sacrament of Holy Communion as an example. The eating of
bread and the drinking of wine are ordinary actions meaningful in
themselves as a means of satisfying hunger and of replenishing
biological life. There is also a further natural result that occurs when
people dine together, when they "break bread" together. When with
one another they eat the same food at the same time and place, they
establish a community of feeling, a conviviality. They become to some
extent—at least for the moment—one. A ceremonial eating and
drinking together can therefore serve as a metaphorical vehicle
embodying (and so "celebrating") the communion of men in a
brotherhood of fellow feeling and can act upon us in a way very similar
to the way in which any metaphor—or possibly any work of art—can
act upon us. We embody our own temporal, inchoate experiences,

stretching over a long period of time, in the unification of one moment—which also embodies our unification *with* one another in the same moment. If this happens to be a traditional ceremony, hallowed by time, engaged in at regular intervals—perhaps by generations of the same people and their descendants—the ceremony will border upon the sacred or religious. Let us say rather that it could move easily into the obviously sacred or religious with the addition, in time, of a myth describing how this ceremony was "once upon a time" ordained by a god.

The presence of such a myth as one of the clues integrated into the meaning of the ceremony detaches the actions performed in the ceremony not only from the ordinary round of prosaic meals—for a nonsacred ceremony would do this—but from the whole realm of temporal events. Through the myth we dwell for the moment in Great Time and are one, not only with one another and with our fathers, but also with the All. We participate in an ultimate meaning of things.

In Holy Communion the myth, of course, is the story of the Last Supper in the upper room in which the Lord himself instituted the rite, to be performed until he should return "in remembrance of me." Added to this meaning, through its mutual embodiment in this myth, is the further metaphoric meaning of the satisfaction of a spiritual hunger and a replenishing of the spiritual life through the ritual assimilation of the body and blood—the substance—of the Son of God, which the bread and wine are. (Somewhat, perhaps, as the Bororos are red parrots.) It is only through belief in the myth, of course, that the whole rich meaning of the ceremony—the congeries of all the sacred, religious meanings—is achieved: all the incompatible meanings that reach their full bloom together in "the New Testament in my blood." We see that this Christian rite is made sacred and detached by the myth that is embodied in it (as it is also embodied in the myth), just as we found that certain primitive rites are made into something wholly detached and sacred by the presence in them of a creation myth.

We have found that works of art are fusions of incompatibles and that this fusion seems to be characteristic of the whole class of meanings we have been considering in this chapter—meanings that we have called "acceptances" as opposed to "observations." We have indicated that religion is one of these fusions of incompatibles with respect, at least, to its rites and ceremonies and myths. Let us now explore more fully the ways in which rites, ceremonies, and myths function in religion as fusions of incompatibles, and then go on to see how the worship involved in religion also functions as such a fusion.

Rites and ceremonies break into our normal routines and introduce an action into our lives (a celebration) that is not an action in the ordinary sense of the word. Our ordinary actions are all located within the temporal frame and are directed at specifics—at specific materials oriented toward specific ends at specific times and places. In fact, *timing*, the choice of the proper time—for the whole action and for each of its parts—is of the very essence of genuine, acceptable action. But the action of a ritual has meaning only in terms of Great Time—the time before all time—which has and needs no date. It is not, except in the case of magic and superstition, aimed at giving effect to specific objectives.

What gives a rite this larger meaning is, of course, the myth that it recreates. The rite is therefore embodied in the myth—which is a curious circumstance, since a myth itself has no body, except in the rites which recreate it. Each exists in a viable form only in the other.

Thus it is not only what is *said* in the myth that serves to detach it from the practical affairs of our lives, but it is perhaps even more the rites and ceremonies that recreate its expressed actions. Each serves as a "frame" for the other's "story." And each frame is, in turn, incompatible with the contents of each story. An action in mundane time and space is framed by one that is outside mundane time and space, and vice versa. Yet they are joined together in a meaningful whole by our imagination.

Consider, again, the Christian Holy Communion. In general, bodily nourishment is thought to interfere to some extent with nourishment of the soul. This apparent fact is what lies behind the universally acknowledged efficacy of fasting for purposes of spiritual edification and progress. Yet the two supposed incompatibles (nourishment of the body and of the soul) are combined in the rite of Holy Communion. There are also obvious incompatibilities involved in considering the same physical objects to be both flesh and bread, both blood and wine—to say nothing of the impossibility of deriving an infinite supply of food from one finite human body. The whole ritual, combined with the myth, bristles with irresolvable incompatibilities. But it is the fusion of these incompatibles, accomplished by our imagination, that gives meaning to the whole transaction and moves our religious feeling so powerfully—if we are Christian.

As we turn our attention to religious worship, we meet with incompatibilities at least the equal of those involved in myth and ritual. How could the infinite God of all Gods, the God of all worlds, the God who "has the whole world in his hand," be in any way pleased, edified, or honored—much less glorified—by the voices and actions, the postures, or even the highest thoughts of a few anthropoidal creatures, only recently descended from the trees, performing rituals in certain finite places, thought by them to be hallowed, and at certain finite times, considered by them to be holy days? The whole "frame" in which the "story" of God's praise and glory is given its location— its embodiment—is ludicrously incompatible with such a "story."

The same sort of incompatibles arise when worship turns from praise and glorification and reverence to thanksgiving. How can it be meaningful to thank, for certain specific blessings received by certain specific persons, the God of all, whose very essence is thought to be always to do what is best for all? But when we inspect a particular instance of rendering thanks to God, these incongruities seem to us to be inconsequential. There is a stone set into a wall at New College, in Oxford, as a memorial to someone long since dead, which reads: "I thank my God for every remembrance of you." This declaration strikes us as a very moving and meaningful action—yet surely only in a wholly transnatural way. There is no logical or practical sense to it. Yet it says something beyond the expression of a sentimental attachment to that person, and the incompatibles that reflection would reveal in such an act of thanks to God are not visible in the meaning into which our imagination fuses them.

Precatory prayers are perhaps, of all parts of religious worship, the most empty of ordinary-action significance. It might be considered remotely meaningful to feel gratitude that God is good—even though we think he could not be otherwise. But how can one presume either to advise or to plead with him to do something good for someone? Such entreaties should logically mean that we do not trust that he will know or will do what is best without our intercession. However, such a prayer, in its most sincere form, is obviously a supreme act of trust. Like the murder on the stage which is actually a nonmurder, this act of nontrust is actually one of trust. The difference is that what on the stage appears as the story is not an action that is actually going on, whereas, in a religious service, what appears not to be truly going on (because it could make no logical sense for it to be going on) *is* actually going on. One *is* revering and honoring God, one *is* being thankful to him, and one *is* trusting him, even though it would appear that one could not possibly accomplish these things by these actions. But the "is doing" here is an "is doing" that passes beyond all modes of temporal action and embodies the truths that make us dwell in Great Time—or, as the Christian puts it, in the Kingdom of Heaven.

It is therefore only through participation in acts of worship—through dwelling in these—that we see God. God is thus not a being whose existence can be established in some logical, scientific, or rational way before we engage in our worship of him. God is a commitment involved in our rites and myths. Through our integrative, imaginative efforts we see him as the focal point that fuses into meaning all the incompatibles involved in the practice of religion. But, as in art—only in a more whole and complete way—God also becomes the integration of all the incompatibles in our own lives.

These incompatibles include not only all the false starts and stops in our lives, the blind alleys, the unfinished things, the loose ends, the incompatible hopes and fears, pains and pleasures, loves and hates, anguishes and elations, the memories, the half-memories, the forgotten moments that meant so much to us at the time, the disjointed "dailiness" of our lives—in a word, all of our inchoate memories and experiences—but also the incompatibles that make up the whole stance of our lives: the hope that we may be able to do or achieve what we know we must do but which we also know we have not the power to do.

In a practical sense these fundamental incompatibles are often resolved by throwing away one or the other. The megalomaniac rejects his frailties, the opportunist rejects his obligations, and the suicide rejects his hope. The sane man, we say, holds all these incompatible factors together in a sort of permanent tension, hoping that somehow he may be given the power to do what he knows he must, but living in the meantime humbly within the limits of his capacities—within his "calling" in the broadest sense of this word. As a matter of fact, this is the sort of faith and hope that a *scientist* has when he faces a problem he does not know how to resolve but which he tackles anyway.[1]

Such a faith and hope are necessarily blind, however, and therefore difficult to maintain under pressure. In the absence of a recognition of the legitimacy of the kinds of imaginative fusions we have been calling "acceptances"—when, in other words, only the observations and the hardheadedness of a scientism or of a "pure" reason are recognized as valid forms of knowledge—it is easy for such faith and hope to be supplanted by the supreme arrogance of a Marxism, which forgets or denies human limitations, or by the freedom of a Sartre, which forgets the obligations we find in our position, or by the despair of a Camus, which abandons all hope as objectively groundless.

In Pauline Christianity, on the other hand, faith and hope have an object. We dwell in the hope that we may, by the grace of God, be able somewhere, somehow, to do that which we must, but which we can at this moment see no way to do—or else trust, if we should never receive that grace, that it is best that we do not do it. Dwelling in this religious frame of mind, we have not lost the tension, but it neither worries us nor do we become complacent. Our myths tell us of the Fall and of how and why we are excluded from the Paradise we long for as our natural state. But they also tell us of the Redemption and of the power and grace of God that is to be dispensed to us as needed. So we are freed from worry about our (to us) insurmountable limitations. But we are not freed from obligation to "the Law," and therefore we cannot become complacent. Rather, we are humbled before God in the recognition of our utter dependence upon him for the ultimate victory through Christ.

None of these beliefs makes any literal sense. They can be destroyed as easily as the actuality of Polonius' death upon the stage, should anyone attempt to defend its reality in the world of facts. Both are works of the imagination, accepted by us as meaningful integrations of quite incompatible clues that move us deeply and help us to pull the scattered droplets of our lives together into a single sea of sublime meaning.

That men have been so moved by their religions surely passes without contention. That some men are so moved today is probably also acceptable, whatever theories may be invented to "account for" it (i.e., to reduce it to naturalistic "causes"). But that it is difficult, if not impossible for many, many men—perhaps for most men—to be so moved today seems also beyond dispute.

But why the acceptance of religion (i.e., of religious meanings) in the sense we have just been describing—not in the sense of religion reduced to a drive for social betterment, for personal character-building, for healthier human relations, or for anything else that is actually something else, but religion as it entails meanings unique to itself—why the acceptance of such a religion should experience such great difficulty in our day: this is the question.

Let us return for a clue to something involved in the acceptance of a work of representational art, something that we took no special notice of in our earlier discussion. The "story" part of such works must have some degree of plausibility. It must strike us, the audience, that a man like Hamlet might kill a man like Polonius under the circumstances presented in the play. Indeed, the whole of *Hamlet* must strike us as some coherent or connected series of events that might issue in the denouement that it does issue in. This requirement is slightly different from the merely cool, "objective" judgment of its logical possibility. As Aristotle pointed out long ago, a convincing impossibility in a play is better than an unconvincing possibility.[2] All the clues, as they reach us, must be such as to induce our imagination to make an integration that seems to be a plausible order of events.

We are used to saying, in our present age of rapid and far-reaching innovations in discoveries, inventions, and the arts, that *anything* is possible. Yet we do honestly and in fact find in anticipation that only certain things are plausible. It is these that we expect and are ready to accept—even if *in fact* they are actual impossibilities. But whatever is plausible to us does, of course, seem possible to us. So let us say that the story part of a work of representational art must seem possible to us in this sense—not as the merely *potentially* possible, which would, for us, include almost anything imaginable, but the *actually* possible, in the sense of what in fact seems plausible to us.

As we have shown, it is sacred myths that give embodiment to the rituals in a religion. But the myths contain a story, and so they have a representational content—even as a work of representational art has. It is this representational content of a religious myth that must seem possible to us (i.e., be actually plausible) if we are to be able to accept it.

It is beyond much doubt that this representational content of the religious myth is at least one of the serious stumbling blocks to the acceptance of religion in our day. So much is this so that a whole school of theologians has become busily engaged in demythologizing our religions. But, if it is true that myths are an essential part of any religion, the success of such a movement can mean only the total demise of religion.

Let us ask, then, what sort of possibility the sacred myths that inform religious rites must have in order to gain our acceptance. We see at once that their possibility cannot lie in our regarding their accounts of events as factually true in the sense of day-to-day possibilities. That is, their possibility cannot lie in our conceiving the events as they represent them as actually having occurred in secular time—at least not *as* such events as these would occur in secular time—because their very detachment rests upon their events being understood as having occurred rather in that "Great Time"—that out-of-this-world time—that Eliade speaks of. If the events in a sacred myth must lack this sort of day-to-day possibility—the possibility that events represented in representational art must have—then, whatever their possibility may be, it must be of a different sort.

Visionary art has shown us that, even when the story content of a work of art quite obviously has *no* plausibility, it is nevertheless possible for our imagination to integrate these incompatible elements into a meaning—a meaning that cannot be expressed in any set of coherent, explicit statements, a meaning that is born and remains at the level of feeling but which is nonetheless a genuinely universal personal meaning and not merely a subjectively personal meaning.

To some extent, perhaps, and for some people, the meanings achieved in religion may be of this same sort. The contents may continue to seem completely implausible to us, while yet we see in the creation stories, the miraculous-birth stories, the Crucifixion and Resurrection stories a meaning expressing the whole significance of life and the universe in genuine and universal feeling terms. Then we can say: It does not matter. If not this story exactly, then *something like this* is somehow true—in fact, is somehow the highest truth about all things.

What then stands in the way of our attaining such meaning in religion today, especially since we do attain it so readily in the visionary arts? When we bring religion and the visionary arts face to face, we begin to see more clearly why the myths and rites of religion (when seen as containing implausible contents) are less easily integrated than visionary works of art. The reason lies in the fact that even when all the representational details in the myths are clearly and frankly regarded as impossible (as the "contents" of visionary art are), the *import* of these details must still be thought to be *plausible*. For, unlike the contents of a work of visionary art, the contents of a religion will have as their import the story of a fundamentally *meaningful* world, whereas the import of a work of visionary art is rather that the world is a meaningless heap of inchoate things. Therefore, if we can regard religious myth as plausible, the sort of world that religious myth represents—a meaningful world—must be thought by us to be plausible. We must be able to say: If not this story exactly, then *something like this* story is how all things are put together. In other words, it must be plausible to us to suppose that the universe is, in the end, meaningful. As William James put it, it must be possible for us to think that the "values" religion says are greatest really *might* "throw the last stone."³ To put it negatively, we must not believe that the universe is such that such "values" *cannot* "throw the last stone," because, if we do believe in such a "value-free" universe, then, as James said, the "religious hypothesis" is not a viable one to us and we cannot entertain it—however meaningful life might become for us if only we *could* entertain it and if only we could thus, perhaps, choose to accept it.

We might even want to go so far as to say that the "religious hypothesis" (to use James's crudely pragmatic term) is so terribly attractive to almost any man that, could he only believe it to be possible, he would alse believe it to be true. The whole experience of mankind has surely been that in general men *do* have such a "will to believe." Some men do believe, even in our day. And for others, as even Sartre, who does not believe, said: We are distressed that God does not exist.⁴

But, however desirable they may be, the meanings of religion will not be likely to be restored to man until his views of the universe are such that he can once more seriously entertain these meanings as representations of the way things could indeed be.

We must now turn to the question whether the naturalistic and scientistic commitments of twentieth-century man irrevocably bar his acceptance of such a possibility, as it appears they seem to him to do, or whether he has merely misconstrued them to stand in his way.

11

ORDER

As we have seen in the last chapter, the representative element in all religious orientations portrays the world as meaningful; that is, it portrays the world as something more than a conglomeration of physical and chemical interactions issuing, to no purpose whatsoever, in whatever ephemeral globs the equilibration of forces renders necessary or probable. For, as many astute thinkers from Socrates on have seen, the world cannot be thought of as *ultimately* meaningful unless the organization of its parts is meaningful, that is, unless there is some point to the way things are put together or, at least, to the direction in which they are developing. This would mean that we would have to attain a view of the world in which the universe, per se, is not "value-free." Some intelligible directional lines must be thought to be operative in it.

Therefore, in order to find the world meaningful, it is insufficient to suppose it to be meaningful only in terms of formulas that explain everything as resulting from orderly and "rational" interactions of forces. Socrates was properly disappointed when he found that Anaxagoras, although he had claimed to show how "mind" was the arranger and cause of all things, had actually shown only that "airs, ethers, and waters" (material elements) were the causes of all things. Socrates then objected that Anaxagoras could not show that things were arranged the way they were because it was *best* for them to be this way. The same difficulties prevail for us today. For we do not hold that the world is absurd because we think the elements in it are not orderly in their relations to one another. We *do* suppose that they are indeed orderly. We think the world is absurd because it seems to us that there is no point to what transpires in it, i.e., that there is no end or aim or purpose to the whole business. It seems to us that there is no meaning to the universe—except possibly the subjective meanings that man tries to import into it. But, in the end, the universe cancels these out. Man thus sees himself to be like a little boy who continually repairs and rebuilds his sand castles at the edge of the sea, only to see the waves continually washing them away. Such fruitless activity is quite acceptable to boys, but men, finally recognizing that the sea sets no store by sand castles, usually suppress their desire to build them—or at least their desire to take their constructions very seriously. If some people continue to build sand castles in so hostile an environment, they will do so only as a diversion—as something to pass the time away—or as a symbol of defiance against the meaninglessness of the universe.

Intellectual assent to the reduction of the world to its atomic elements acting blindly in terms of equilibrations of forces, an assent that has gradually come to prevail since the birth of modern science, has made any sort of teleological view of the cosmos seem unscientific and woolgathering to us. And it is this assent, more than any other one intellectual factor, that has set science and religion (in all but its most frothy forms) in opposition to each other in the contemporary mind.

The recent rise of philosophical (and popular) opposition to science has not, however, reinstated teleology, for this opposition has taken forms that are as fully opposed to the notion of any sort of cosmic purpose as contemporary science is. Existentialism is rooted in an antideterminism that asserts a radical freedom (on the part of human beings), and it therefore also tends to abhor teleological views, not on scientific grounds, but simply as another form of determinism.

It is true that the teleology rejected in our day is understood as an overriding cosmic purpose necessitating all the structures and occurrences in the universe in order to accomplish itself. This form of teleology is indeed a form of determinism—perhaps even a tighter form of determinism than is provided for by a materialistic, mechanistic atomism. However, since at least the time of Charles Saunders Peirce and William James a looser view of teleology has been offered to us—one that would make it possible for us to suppose that some sort of intelligible directional tendencies may be operative in the world without our having to suppose that they *determine* all things.[1] Actually it is possible that even Plato did not suppose that his

"Good" *forced* itself upon all things. As Whitehead has pointed out, Plato tells us that the Demiurge, looking toward the Good, "persuades" an essentially free matter to structure itself, to some extent, in imitation of the Forms. Plato appeared to Whitehead to have modeled the cosmos on a struggle to achieve the Good in the somewhat recalcitrant media of space and time and matter, a struggle well known to all souls with purposes and ends and aims. Whether or not it is true that Plato did this, certainly Whitehead modeled his *own* cosmos very much this way.[2]

The teleological orientations of things in the world that such philosophers as these have envisaged, rather than substituting a determinism of purpose for one of forces, have held that the tendency toward achievement of more meaningful or orderly or regular relations (which they thought was clearly exhibited in the world) was inexplicable unless some principle operative in that direction was immanent in all things. As Whitehead put it, there is too much order in the world to be explained without recourse to such a principle.

Yet, in spite of the fact that our contemporary minds have been offered a few first-rate teleological alternatives to the choice between blind mechanical necessity and total freedom, we in the modern world seem on the whole to have remained committed to one or the other of these opposed polarities and thus to the notion that the world, as such, is simply absurd. "Teleology" has therefore become a dirty word to us. Before Darwin's mechanical explanation of the origin of species through natural selection and before the immense strides of modern biology in providing us with mechanical (chemical and physical) explanations of life processes, teleology had not become quite such an objectionable notion to us. Living organisms, at least, seemed to be purposeful in their organization (an integrated structure of functioning organs and tissues) and in their operations and their ecology as well. But now the reduction of living organisms—without residue—to physical and chemical interactions seems to be taken for granted in biology, both as a project to be accomplished and as a fact that only fuzzy-minded persons would ever doubt. It is but one step from this position to a completely behavioristic approach to animal and human psychology, to one that wishes to abandon altogether all "mentalistic" talk of purposes or aims or goals in the discussion of animal behavior (including even the "highest" human behavior).

Therefore, assuming that the current conviction that life can be reduced to physical and chemical laws is the chief stumbling block to the possibility of our entertaining any sort of teleological views, let us survey in very general terms the state of our present scientific knowledge of living organisms and of life.

We believe that life began (on earth, at least) a few thousand million years ago and that it contained from the start long compounds of DNA, a linear sequence involving four chemical radicals. We think that the particular arrangement of these four chemical radicals conveys a vast system of notations, different for each kind of organism, which is transmitted chemically through successive generations.

Twenty years ago Watson and Crick identified the structure of this chain, and inquiries made since then have found that all members of the entire system of terrestrial living beings are carrying the same system of notations arranged in their own peculiar alignments. The entire complex of terrestrial life is based on chains of the same four DNA compounds, each organism having its own distinctive sequence of them.

These sequences of DNA have conveyed from generation to generation the hereditary features of living things, all the way from the beginning of life on earth to the present. The DNA chain of each organism is transmitted from living beings to their descendants by subdivisions which are linear sequences for the lower monolithic organisms and binary combinations of such strains from opposing sexual organs for bisexual organisms. In both cases the characteristic sequence of the parental bodies is known to be transmitted to the new organism, with such variations as may have arisen from mutations, but further amplified by innovations due to the combination of sexual pairs in bisexual organisms. Hence it is possible for the innate pattern of each newborn organism to be novel to some significant extent.

Each such new pattern conveys the characteristic pattern of the parent, or parents, to the offspring with such modifications as flow from parental innovations, including the effects of biparental combinations. This is how successive features are shaped in the transmitted patterns and a continuous flow of hereditary innovations is sustained. These continually occurring random changes offer the occasion for the development of new species by natural selection; those species able to survive under the circumstances thus pass their innovations on to further generations.

So the pattern of the DNA chain both sustains the pattern of the current organism and offers ever renewed chances for Darwinian innovations which lead to the descent of new, higher organisms from originally lower forms of them. This process establishes a measure of growth in DNA sequences, as their DNA becomes increasingly extended in the developing evolution of species. This increase is measurable in the DNA chain by the increase in the number of possible variables. The DNA chain of a bacterium carries about twenty million alternatives; an insect has ten times more, i.e., two hundred million; and a human being has about twelve billion in his DNA chain, about a million times the number possessed by a bacterium. We human beings are therefore thought to be a million times more complicated than bacteria.

This pattern of initiating and developing living beings is so amazingly uniform in all plants and animals that it has caused informed people to think that it constitutes the key to the final explanation of all life in physical and chemical terms. But this is far from the truth. The chemical composition of DNA is astonishingly simple, and its sustenance and renewal of all living beings is so comprehensive that it does indeed invite the supposition that it has explained the main features of life in chemical terms. But great parts of this secret are missing.

We are faced first with the question how the inanimate realm of the earth, from which all life has arisen, ever produced a sample of a DNA compound. A number of calculations estimating the probability that such a synthesis will take place accidentally have agreed in finding the chances of this so small, so rare, that the event appears virtually impossible.

But suppose we set this difficulty aside. How then can we account for the transmission of life from one generation to the next? A fertilized egg will develop into an organized system including thousands of millions of new cells in the modified image of its ancestors. The successive stages of the embryo are each larger than its immediate precursor, each more differentiated, and all these transformations are accomplished by the division of existing arrays of cells into twice as many new cells.

Each step in this expansion is induced by the surroundings of the DNA compounds at that stage, and these guiding surroundings are due to the guiding surroundings of a previous stage, and so on. In other words, each step of the biological growth anticipates the subsequent step of growth by producing for its guidance a cellular body which will instruct the cellular divisions of the next stage and will carry on the development to the final product.

The power of DNA to guide the physiological development of the fetus has been compared with the power of an architectural blueprint to guide the construction of a new edifice. But its function is far more complex. A given DNA molecule guides the successive stages of embryonic development; each stage produces a cellular milieu which will in turn guide this DNA in producing a further cellular milieu for very *different* actions of this same DNA—those which have to follow next in order to complete the proper development of the embryo. The *same* DNA particle, in other words, acts differently at different times in the different stages of the development of the embryo. A blueprint does not act differently at different times in the construction of a building. In fact, of course, it does not act at all. It merely provides a guide for the builder so that he can vary his behavior in a purposeful way as the building progresses, assuming that he aims at producing the finished structure anticipated in the blueprint.

Since the chemical compound DNA is assumed to act only chemically, it cannot vary its actions in the way a builder with a mind can. Therefore, there must either be another element in the organism that can function as a builder, merely using the DNA compound as its blueprint, or, as the theory supposes, the DNA must merely be responding chemically to chemical compounds. In the latter case, if it acts differently at different times, there must be different chemical compounds for it to react with at different stages of embryonic development. But these compounds must be called into existence only *at the end* of the embryo's previous stage. Timing is therefore most important. No theory yet exists to explain how this can be done in a strictly chemical way. The development of such a theory is rendered more difficult because, as Driesch first showed in his work with sea-urchin embryos, there seems to be some resilience possible in the development of tissues. Some tissues can sometimes seemingly be "pressed into" undergoing changes they do not normally undergo, because some part of the embryo has been prevented from developing in its usual way. It is almost, in these cases, as if there *were* a builder who has had to use some ingenuity because of shortages of one material or another or because something has previously been built erroneously and he must now build upon and around this.

All these considerations seem to point to one of two conclusions: either the DNA is at once the blueprint and the builder (it is a sort of "master molecule," and it makes adaptations in some kind of purposeful way), or else it functions as merely another "organ" in the body and so is interrelated in an immensely complicated way with every other organ (and *cell*) in adapting itself to the needs of that organism for growth and maintenance.

That there is a finite number of physical and chemical mechanisms available in any given DNA molecule for making adaptations to different circumstances (due simply to its chemical structure) and that these mechanisms are triggered merely by the physical and chemical conditions that prevail when there is such a "need" for such adaptations may indeed be true, and the existence of these mechanisms may remain undiscovered until they show up under some unusual circumstances. But since we are unable, from the structure of the DNA, to predict their existence chemically, we must admit that we do not yet have the reduction of living processes to physical and chemical laws that modern biologists seem to think we can have. Moreover, if in fact a truly new mechanism *were developed* by the organism in response to a new "need," we would still be faced with the *choice* of supposing either that it was truly the creation of a genuinely new response to a new situation in its embryonic development or that we have for the first time seen a built-in mechanism that was really there all the time but had had no occasion to come into play before. In other words, there seems to be no way such a question as this could ever be firmly and objectively answered. We not only *have not proved* that these adaptive aspects of the DNA's building capacity can be reduced wholly to physical and chemical operations, *but we never can do so*.

Another unsolved problem arises from the continuous quantitative increase in DNA chains from those of bacteria to those of man—from about twenty million DNA alternatives to about twelve billion. There is no chemical model available to explain this enormous growth of the DNA chain from those in bacteria to those in man. We have no chemical explanation for this fundamental fact of the system, just as we have no chemical explanation for the historical origin of DNA or for its capacity to produce media that apparently anticipate the continued development of the embryo.

The discovery of DNA and its derivative processes has indeed thrown a sharp light on the structure of living things, but it has also posed these deeper questions for us. Certainly we must say that, in terms of our present scientific knowledge, the biological grounds of our existence are by no means entirely clear to us. What alone is clear is that these grounds have not yet been carried back entirely to the laws of inanimate nature.

So far we have shown not much more than that completely chemical and physical explanations for certain crucial aspects of living things have not yet been found, in spite of the momentous discovery of DNA. What we now wish to do is to go beyond this observation and exhibit why it is we think that such explanations *cannot* be found.

It has for some time been generally supposed that organisms are mechanisms and that, since mechanisms work in accordance with physical and chemical laws, organisms must also do so. This has been a very fruitful assumption in terms of the discoveries of physiological mechanisms that it has evoked. Its fruitfulness rests, of course, on the guidance and the impetus it has provided for research into the physical and chemical interactions that go on in a living organism. Unfortunately this assumption has been misconceived to mean that organisms must be wholly explicable as the resultants of physical and chemical laws because mechanisms are. Actually it means exactly the opposite. For mechanisms are *not* wholly explicable as the resultants of the operation of physical and chemical laws.

There are two kinds of principles involved in the very conception of a mechanism. One kind is, of course, the physical and chemical reactions that occur "necessarily," as we say, given certain physical and chemical conditions. These are what make a mechanism "work." And unless we could rely on them to take place, we could never in any way predict what a mechanism would do—which would only be another way of saying we would not have mechanisms. There simply could not be any.

But another kind of principle essential to mechanisms is that which sets the boundary conditions—the limits—within which these *particular* physical and chemical interactions take place. These boundary conditions determine the *structure* of the machine—the peculiar organization of its physical and chemical parts. What they determine is not the ways in which these parts do interact with one another, once they are structured in this way and once the key parts are supplied with sufficient energy, but rather the *pattern* in which they are put together. This organization of the machine's parts is not a resultant of the operation of physical and chemical laws. Machines made by man are not the random resultants of a physical and chemical equilibration of forces that just happens to take place. Each machine has been the result of man's efforts guided by his knowledge and imagination and directed toward something that such a machine is to *achieve*.

Thus we cannot continue to think of organisms as mechanisms if we wish to explain them in terms of *one kind* of principle only, namely, the physicochemical kind. Our explanation would be lacking the boundary-type principle that an understanding of something as a mechanism requires. But the notion that organisms are, in fact, very complexly organized systems of mechanisms has been, as we have said, a very fruitful notion, accounting for practically all of the immense growth of biological knowledge in our day, including the DNA breakthrough. It would seem foolish, therefore, for us to abandon the assumption that organisms are mechanisms, and apparently no one in biology is actually doing so.

One thing obvious about mechanisms is that each one of them has acquired its organization by reference to some aim or goal or purpose that is to be achieved by it. That this purpose is not something deducible from the physical and chemical laws that operate its parts can easily be seen by the fact that, although we may be very well versed in these laws, we cannot, through the mere physical and chemical study of even simple machines, such as manual tools, tell what the tool is *for*. If we do not know what it is for from other sources, such as from observation of the tool or machine in use, we remain in ignorance of it. What some presumed artifacts from former civilizations were used for remains a mystery to us, although their physical and chemical composition is no secret.

It might be claimed, however, that organisms (which, of course, have not been made by man) have simply resulted entirely from the operation of physical and chemical laws (even though we at present have, as we have seen, no adequate theories for explaining some very basic problems concerning how this could have occurred) and that they thus do not have this dual origin, that they only *seem* to us to be functionally organized in their parts and, as a whole, to accomplish certain things. If living things do present us with illusions, these are indeed grand illusions. For living organisms do certainly seem to us—even to research biologists—to be functionally organized the way mechanisms are, i.e., by meaningful boundary conditions. The illusion that they are so organized is so great that they readily fool us into thinking that they are even distinguishable from various inanimate processes that closely resemble them, processes such as open systems. Open systems, such as flames and thunderstorms, are known physical and chemical processes that sustain themselves in existence for a time, "feeding" upon what they "produce," growing, and in some sense reproducing. Yet it seems quite clear to everyone, including even biologists who wish to reduce living organisms wholly to physical and chemical transactions, that such open systems are not living organisms—that there is something decidedly different in their modes of operation.[3]

Just what this difference is seems to be difficult for biologists to say definitively. Their difficulty may be related to the fact that we appear to require the category of "achievement" in order to talk sensibly about living things. We have a genuine need, for instance, for a science of pathology in connection with them. We know that organisms may succeed or fail; and unless we know the causes of their failures—their malfunctionings and their diseases—we do not know all there is to know about them. A proposal for such a science in connection with thunderstorms would not, however, be appropriate, and the reason is that the category of achievement has no place in our understanding of thunderstorms. Their parts do not appear to us to be organized to perform certain functions essential to the maintenance of thunderstorms—even though they *do*, in fact, maintain them. Nothing gives us the appearance of "trying" to achieve a thunderstorm. Certainly the thunderstorm does not; its behavior seems to lack the integrity of an individual. But even a paramecium is an individual that quite apparently strives (whether consciously or unconsciously makes no difference) to adapt itself to its conditions and to stay alive and to reproduce. It definitely possesses adaptive mechanisms for the achievement of these "ends," and these mechanisms may succeed or fail in any instance. By contrast, hydrochloric acid can never fail to dissolve zinc. Nor can it ever dissolve platinum by mistake. Only living things can make mistakes. Only living things can fail—or succeed.

There is little to be gained here from further arguing this point. If someone insists that even our best biologists today are simply being "animistic" (it would indeed be ironical, of course, if he should use this term) or superstitious or mythological when they take this attitude toward living things and that—appearances notwithstanding—there is *no* difference in the *basic* explanatory principles required for dealing with a thunderstorm as against an elephant, little more can be said to him that is directly to the point. He would appear to be maintaining a thesis at all costs, for he is holding gratuitously that a principle of distinction we all do use, because observed facts seem to call for it, is not in reality necessary. At best he is using a blank check to draw on the future, counting on deposits that he presumes reductionistic biologists will one day make.

However, there are other considerations that indirectly cast doubt on this die-hard position. DNA has been said to function as a code for the chemical and physical development of the organism—as a sort of chemical message or messenger. Let us examine the nature of codes and messages. It is clear that nothing can function as the carrier of a code unless it is chemically and physically neutral to the messages it is to transmit. Any medium fails as the medium for a message to the extent that merely random physical and chemical equilibrations of the sort employed for recording the message can impress themselves upon it. Such random impressions will register as mere noise and will interfere with the message. The static mixed with radio signals, the hum of a radio receiver, the electronic imperfections of an electronic recording tape all mar their messages and, if loud enough, obliterate them entirely.

Blobs and pulsations can function as a message only if they are *not* the result of merely random physical and chemical equilibrations and affinities. It is because rocks do not arrange themselves in the shapes of letters by means of the operation of random physical forces that they can be used to spell out the name of a place on an embankment. It is because ink blobs do not have a physical or chemical affinity for paper *in the particular shape of language symbols*, and so do not arrive upon pages in these shapes through merely random equilibrations of forces, that the ink-blotted pages conveying this present message *can* convey this present message. And it is because the items in a DNA chain do not *have* to be arranged (chemically) in the manner in which they are that it can function as the medium for a message. Each item of a DNA series consists of one of four alternative organic bases (actually two positions of two different compound organic bases). Ideally, in order to convey the maximum amount of information, these four organic bases should have equal probability of forming any particular item in the series. Actually DNA does fall below the ideal as a medium because there is *some* difference in the probability of the bindings of the four alternative bases, i.e., there is some redundancy; but it is not enough to prevent DNA from functioning effectively as a code. If DNA were an ordinary chemical molecule, its structure would be due to the fact that it had achieved maximum stability in that structure, and this chemical orderliness would prevent it from functioning as a code. But it is not an ordinary molecule, and so it can function as a code.

It should be interesting to note here that whether or not this codelike structure of the DNA molecule is assumed to have come about by a sequence of chance variations, some of which have become established through natural selection, the point we are making holds true. Whatever the origin of a DNA configuration may have been, it can function as a code only if its order is not due to the forces of potential energy. Just as the arrangement of a printed page is and must be extraneous to the chemistry of the printed page, so the base sequence in a DNA molecule is and must be extraneous to the chemical forces at work in the DNA molecule. This physical indeterminacy of the sequence of items in a DNA molecule makes for the improbability of the occurrence of any particular sequence and thereby enables a sequence to have a meaning with a potential information content equal to the numerical improbability of any particular arrangement coming about merely through the minimization of potential energy.

Modern theory (wishing to confine itself to physical and chemical forces and modes of operation) is left with nothing but chance as the originator of the particular structures of all the DNA molecules. Since anything can happen by chance—even the most improbable thing—these molecules may have happened into being by pure chance. But it would seem to be highly improbable. The chances that merely *by* chance they should have become arranged in the meaningful ways in which they are arranged are beyond much doubt less than the chances that a pile of rocks rolling down a hillside will arrange themselves by chance on a railway embankment in such a way as to spell out in English the name of the town where the embankment is.

The fact is, therefore, that every living organism is a meaningful organization of meaningless matter and that it is very highly improbable that these meaningful organizations should all have occurred entirely by chance. Moreover, looking at the general direction the evolutionary development of living organisms has taken, one must in all fairness admit that this direction has been toward more meaningful organizations—more meaningful both in their own structure and in terms of the meanings they are able to achieve. From microscopic one-celled plants, able to do very little more than to provide for their own sustenance and to reproduce, to minute animals, sensitive *as individuals* to their surroundings and able to learn very rudimentary sustaining habits, to more complex animals, able to do many more things, to the higher mammals, and finally to man, who is able to achieve so many things that he frequently supposes himself to be a god able to achieve *all* things—this evolutionary history is a panorama of meaningful achievement of almost breathtaking proportions. And when one considers, in the face of the staggering number of merely chemical possibilities open to DNA, that so many meaningful combinations have come into existence and that these have been oriented in the direction of the achievement of more and more meaning (even when change in this direction did not always relate to the necessities of survival), it is difficult to avoid the notions that some sort of gradient of meaning is operative in evolution in addition to purely accidental mutation and plain natural selection and that this gradient somehow evokes ever more meaningful organizations (i.e., boundary conditions) of matter.

The inclusion of such a gradient in our biological theories should not, therefore, meet with the adamant resistance it does—and especially not when we reflect that a principle analagous to it has been operative in modern physics for many years. In mechanics and thermodynamics and also in open systems we suppose inanimate nature to be controlled by forces that draw matter toward stabler configurations. Current physical theories are quite obviously guided and structured by the assumption of the existence in nature of such a gradient. It is now so much a part of our thinking in physics that we take no special note of it. It seems natural to us that nature consists of forces and that these should establish a stable configuration when they reach a kind of balance—when their potential for change, their potential energy, is at the minimal level possible under the circumstances. But there was a time (before the advent of modern physical science) when such an assumption played no part in physical speculations. It was not then assumed that nature acted for this sort of "end" (which is only to say, acted in such a direction); rather it was assumed that it acted for other "ends," i.e., in other directions, different for different sorts of things and supposedly appropriate to their "natures." Stones fell because their "end"—in the sense of their directional gradient, not, to be sure, their consciously held goal—was downward. Flames rose, on the contrary, because their "end" was to reach greater and greater heights, etc. The ancient and the modern explanations can both be understood in ways that are true to experience. As a matter of fact, the old way, Aristotle's, is easier to square with our ordinary experience of gross objects than modern physics is. This is not because our modern explanations are less accurate; they simply turn out to be more complex, more roundabout, than the ancients' explanations. To explain why a rock falls, *we* have to make a journey all the way to the laws of motion and gravitation and back.

Our modern assumptions that inanimate nature moves in the direction of an equilibration of forces actually substitutes a new sort of "end" in nature for the old "ends." It does not eliminate the notion of "end" altogether if, by "end," we mean simply a directional gradient exhibited by a process. This fact is generally hidden from us, perhaps because we seem to assume that existing forces, in simply acting, *force* a state of equilibrium into existence by mutually limiting one another.

But one of the early modern philosophers who recognized what the new science of his day was about (possibly because he was conscious of how men's ideas were changing in his day) saw very clearly that the new laws of motion functioned as part of the structure of a "final cause." Leibniz noted that the principles of the conservation of force and of equal reaction were not logically or mathematically necessary to the concept of motion or of matter. They rather, he said, require us to make use of the notion of a "final cause," since they depend "upon the principle of fitness."[4]

Regardless of whether one interprets the "fitness" of the laws of motion (or, for us, the equilibration of forces) as, in Leibniz's words, a "final cause," that is, as a directional gradient that exerts its influence upon situations, it remains true that we espouse such a principle in modern physics, not because we see its logical entailment in the concepts of matter or energy, but rather because we believe—contrary to the ancients—that physical events *do* move in this direction. We "believe" this rather than "prove" it from careful observation; for, as to the supposed evidence for this principle in the *observed* balance of the quantity of forces entering into situations, our use of such a directional gradient as this in our analyses of situations provides the basis for our assigning quantitative values in the first place to the "forces" we believe to be involved. We can measure one force only in terms of another; but what is often forgotten is that it is only because we assume they must all balance that we can possibly establish the *quantity* of one in terms of another. We must *assume* an equation before we are able to determine the values of any of the variables involved. We thus have no basis for the measurement of any "forces"—independent of the use of this principle—and their supposed balance is therefore assumed, or defined, by this principle. The principle itself is certainly never *proved*. We believe it to be true, however, because on the whole it seems more "fit" to us than alternative principles for making sense of events in the inanimate realm.

Let us look now more closely at how this principle functions in quantum mechanics. Quantum mechanics (1) assumes a field of forces (energy) driving with certain degrees of probability toward stabler potentialities. This assumption, as we have seen, entails a notion of the existence in inanimate nature of a gradient in the direction of the minimization of potential energy. Quantum mechanics also assumes (2) that these potentialities are not always being actualized, because the forces involved may be friction-locked, and also (3) that catalysts or accidental releasers may come into the situation and release some of these potentialities from their "locked" state, i.e., allow them to become actualized; an explosion, for instance, may be "triggered" by many things, including the spontaneous disintegration of a single atom. Such a spontaneous event may therefore be treated as an *uncaused* event, controlled by probable tendencies but not necessitated (not "caused") by these tendencies. Neither these probable tendencies nor the gradient of the minimization of potential energy could be said to *cause* the ensuing event, although they might be said to "evoke" it.

In an analogous fashion, the growth of an embryo may be evoked (and so also controlled) by a similar gradient of potential *shapes* in a field of shapes, much as the motion of a heavy body down an inclined plane may be evoked (and so also controlled) by the gradient of the minimization of potential energy. A considerable amount of experimental work, done by such biologists as H. Spemann, Paul Weiss, and

C. H. Waddington, has shown that some of the development that takes place in embryos is controlled by fields, although exactly how this occurs is still uncertain.[5] Most biologists would still attempt to seek explanations of this control wholly in terms of a simple single-level ontology, i.e., purely in terms of physical and chemical interactions. The notion of a field in embryology could, however, if it were not handled too mechanically, provide an opening for the entry into the picture of a notion of a hierarchy of levels of being, in which higher levels emerge into existence in and through the establishment of new boundary conditions which in turn reorganize elements of the lower levels in which they are rooted. There is some evidence that some scientists may be making moves in this direction.[6]

The notion of a gradient sloping in the direction of the minimization of potential energy, as we have spelled it out above, can also be used as a model for describing the efforts of human thought. When one sees a problem and undertakes its pursuit, one sees a range of potentialities for meaning which one thinks are accessible. Heuristic tension in a mind seems therefore to be generated much as kinetic energy in physics is generated by the accessibility of stabler configurations. The tension in a mind, however, seems by contrast to be deliberate. A mind responds in a *striving* manner to comprehend that which it believes to be comprehendable but which it does not yet comprehend. Its choices in these efforts are therefore hazardous, not "determined." Nevertheless, they are not made at random. They are controlled (as they are evoked) by the pursuit of their intention. These choices resemble quantum-mechanical events in that they are guided by a field which nevertheless leaves them indeterminate. They are therefore also "uncaused," in the sense that there is nothing within the possible range of our knowledge that determines or necessitates that they become precisely what they do become.

Although discoveries are, in these features, like inanimate events, they differ decidedly, of course, from inanimate events in that the field that evokes and controls or guides them is not a field of more stable configurations of forces but a *problem*. Moreover, discoveries do not occur in a merely spontaneous way but are due to an *effort* to actualize certain hidden potentialities; and the uncaused action that releases, and so also evokes them, is not a physical event but an *imaginative* thrust toward such a discovery.

Let us now show how we can describe the development of living organisms by making use of the same general conceptual framework that we have just seen to be operative in both quantum mechanics and problem-solving.

Both the embryological and the general evolutionary development of living organisms can be understood (1) to be evoked by the accessibility of higher levels of stable meaning. But (2) the attainment of these potentially greater meanings may be blocked by the locked-in code of a given DNA molecule. Such accessibility, analogous to the accessibility of a discovery or of a more stable configuration of potential energy, can be understood, taken together with the existing lock-in of its present structural meaning, to generate a tension that (3) may be released into further development by accident (such as by a mutation-causing event in evolutionary development) or by the operation of releasers or inhibitors (in embryological development), or through the occurrence of first (uncaused) causes—true spontaneities, analogous to the breakup of an atom. A creative development so released could then be said to be evoked, and so also controlled, but never fully *determined* by its accessible meaning potentialities and/or by its releasing agent. Such developments may also, like the thrusts at discovery, succeed or fail to reach a stabler meaning, i.e., a meaning more whole and complete in terms of meaning. The dimension of success or failure is, of course, an *emergent* feature added to the framework of what goes on at the physical level. A stabler configuration of forces is not something these forces strive to achieve and may fail to attain. It is simply the terminal point understood by us to be reached in fact by physical processes.

Such a view of embryological and evolutionary development is not only possible, therefore, but it would also be a view superior to present biological views, since it would succeed in accounting for the rise of living organisms just at the points where the attempts to account for them entirely on the level of the inanimate fail: (1) the boundary conditions, which must consist of principles other than those of the material they bound, would in this view acquire these new principles through a creative reorganization of the DNA chain of an existing organism in response to the gradient of deepening meaning; and (2) we should no longer be faced with the hopeless task of attempting to explain sentience (an obvious characteristic of life, at least in higher animals) in terms belonging to the insentient. Sentience would now be understood to be a structural feature of higher boundary principles of organization and operation, rooted in and dependent for its existence upon a lower, insentient level, to be sure, but *added to* those principles which structure that lower, insentient level—just as certain principles peculiar to calculus, although they are rooted in the principles of arithmetic, are *added to* them, not derived from them, or as the principles of grammar, although they are rooted in the principles governing words, are *added to* them, not deduced from them. Just as arithmetic is not logically rich enough for us to deduce a calculus from it, and words are not rich enough for us to deduce a grammar from them, so the laws of physics and chemistry would now be understood to be not rich enough for us to deduce the characteristics of sentience from them, and we would cease trying to do so. We would instead simply derive the principles of sentience, and those of the still higher states of consciousness, resting for their existence upon sentience, directly from a study of the operation of these states themselves, in a manner unembarrassed and unencumbered by an ideological necessity to reduce them all to levels of being lower than themselves. These principles would then also be available for use, in addition to the physical and chemical principles, in constructing a richer and fuller explanation of the behavior of living things.

Bringing together, for the general problem of this chapter, the implications of what we have been discussing, we find that progress in science is explicable in this view as a development evoked by a gradient of meaning operative in a field of potential meanings and problems. We are thus able to think that real discovery in science is possible for us because we are guided by an intuition of a more meaningful organization of our knowledge of nature provided by the slope of deepening meaning in the whole field of potential meanings surrounding us. We are thus able to know (in some anticipatory, intuitive sense) enough of what we do not know as yet in any explicit sense (because we have not yet discovered it) to enable us to locate a good problem and to begin to take groping but effective steps toward its solution.

We can now understand scientific inquiry to be a thrust of our mind toward a more and more meaningful integration of clues. We have seen that this is also what perception is. We have just shown that living things, individually and in general, are also oriented toward meaning, and it is clear, from our immediately preceding chapters, that man's whole cultural framework, including his symbols, his language arts, his fine arts, his rites, his celebrations, and his religions, constitutes a vast complex of efforts—on the whole, successful—at achieving every kind of meaning.

We might justifiably claim, therefore, that everything we know is *full* of meaning, is not absurd at all, although we can sometimes fail to grasp these meanings and fall into absurdities. In other words, meaning can be missed, since the emergence of life opens up the possibility of success but also, of course, the chance of failure. Moreover, as we have seen, we can claim all this with an open and clear scientific conscience. The religious hypothesis, if it does indeed hold that the world is meaningful rather than absurd, *is* therefore a viable hypothesis for us. There is no scientific reason why we cannot believe it.

But to find no scientific reason why we cannot believe it is not to believe it, especially since, as George Santayana once pointed out, it is as impossible to be religious without having *a* religion as it is to speak without having *a* language.[7] One cannot speak of dogs without having a word for them. The particular sound or sign "dog" might indeed be recognized as merely conventional and of only historical origin; nevertheless, one would have to have either *this* word or else some other equally accidentally established word, and one would have to be committed, in using it, to the notion that it really meant that particular animal. In the same way, we cannot worship, entreat, thank, or praise God (however he is conceived) without having some particular way to do it that seems to us to be set apart and sacred—some way detached from our ordinary transactions with our fellow human beings. Religious meaning, as we have seen before, is a transnatural integration of incompatible clues and is achieved through our dwelling in various rituals and ceremonies informed by myths. These must, of course, be *specific* rites and myths—not just rites and myths in general. There are no such things. Religion "in general" is thus not religion, just as language "in general" is not language. To be religious we must have *a* religion.

In former days one usually inherited one's religion, as one did one's language, one's customs, one's name, and even sometimes one's trade. But a great many people today have not inherited a religion. There is another way, of course, by which we may come to a religion. This is by conversion. But conversion seems to come "out of the blue." It seems clear that we do not become converted—whether to a political party, a philosophy, *or* a religion—by having the truth of what we become converted to demonstrated to us in a wholly logical or objective way. Rather, what happens when we become converted is that we see at some point that the particular party or religion or epistemology or world view (or even scientific theory) in front of us holds possibilities for the attainment of richer meanings than the one we have been getting along with. At that moment we *are* converted, whether we have ever willed it or not; for, as we have seen, we are addressed by nature to the attainment of meaning, and what genuinely seems to us to open the doors to greater meaning is what we can only verbally refuse to believe. As Santayana also said, should we ever "hear the summons of a liturgical religion calling to us: *Sursum corda*, Lift up your hearts, we might sincerely answer, *Habemus ad Dominum*, Our hearts by nature are addressed to the Lord."[8]

William James, good pragmatist that he was, unfortunately tried to render his readers open to the call of a religion in a way that has properly offended many truly religious souls. He held that should our world view be such that the religious hypothesis could be a viable one to us, we could then believe it, deliberately, and try to see whether or not we were "better off," as the religious hypothesis predicted we would be. Try believing in God, he seemed to advise, and see whether or not you feel better about everything.[9]

The truth of the matter is that we may not feel better, or be "better off," if we embrace a religion. We may instead reap suffering, struggle, and sacrifice. Anyway, we do not accept a religion because it offers us certain rewards. The only thing that a religion can offer us is to be just what it, in itself, *is*: a greater meaning in ourselves, in our lives, and in our grasp of the nature of all things. James's conditionally undertaken belief cannot be a genuine belief, since we entertain it with our fingers crossed. In reality, as we have seen, a religion exists for us only if, like a piece of poetry, it carries us away. It is not in any sense a "hypothesis."

It is for this reason that this present work is not directed toward effecting conversions to any religion. At the most, it is directed toward unstopping our ears so that we may hear a liturgical summons should one ever come our way. As Saint Augustine viewed it, a religious belief cannot be achieved by our deliberate efforts and choice. It is a gift of God and may remain inexplicably denied to some of us.

We have in this chapter, therefore, tried only to show that our modern science cannot properly be understood to tell us that the world is meaningless and pointless, that it is absurd. The supposition that it is absurd is a modern myth, created imaginatively from the clues produced by a profound misunderstanding of what science and knowledge are and what they require, a misunderstanding spawned by positivistic leftovers in our thinking and by allegiance to the false ideal of objectivity from which we have been unable to shake ourselves quite free. These are the stoppages in our ears that we must pull out if we are ever once more to experience the full range of meanings possible to man.

But we must not only unstop our ears so that we may hear our god speak to us, should he deign to do so; we must also seek to live in a kind of society in which such meanings as we have been exploring in this work are acknowledged to be real and worthy of respect and honor—and in which men are therefore also respected and honored as creators and bearers of such meanings. If we cannot live in such a society, we shall find ourselves engaged collectively in the brutal task of stamping out these meanings on the grounds of some supposed social utility or for some overpowering ''cause''—a frightful enterprise that modern man has undertaken (and is still undertaking) in this present century.

Let us therefore in our final two chapters show what basic principles must structure a society open to the pursuit of such meanings.

12

MUTUAL AUTHORITY

I̲F̲ ̲W̲E̲ ̲B̲E̲L̲I̲E̲V̲E̲ ̲I̲N̲ ̲T̲H̲E̲ ̲E̲X̲I̲S̲T̲E̲N̲C̲E̲ ̲O̲F̲ ̲A̲ ̲G̲E̲N̲E̲R̲A̲L̲ ̲M̲O̲V̲E̲M̲E̲N̲T̲ ̲T̲O̲W̲A̲R̲D̲ the attainment of meaning in the universe, then we will not regard any of the kinds of meaning achieved by men to be merely subjective or private—to be only a sort of epiphenomenal foam floating to the surface, a mere byproduct of the real interactions that go on in the world, as the Marxists have taught. We will, on the contrary, regard every achievement of any sort of meaning as the epitome of reality, for we will think it is the sort of thing that the world is organized to bring about. However, those of us who cannot believe that one of the structuring principles in the world is a gradient of meaning may still be brave enough to hold fast to all the meanings we have been able to achieve, regarding them as the most precious things we possess. But, whether or not we think such meanings are part of what the whole universe is about, we will, no doubt, if we truly value them, wish to live in a society in which their attainment is honored and respected. We will not want to see any of these meanings demeaned. We will not, for instance, want to see them controlled in some supposed public interest.

It is almost axiomatic that the distinction between a free and a totalitarian society lies exactly at this point: a free society is regarded as one that does *not* engage, on principle, in attempting to control what people find meaningful, and a totalitarian society is regarded as one that does, on principle, attempt such control. What happened to the realm of meanings in the two strongest totalitarian states of our century—the states of Hitler and Stalin—is evidence for the justness of this distinction. Not only was "brainwashing" practiced on a massive scale by the central authorities in these two states through blanket forms of propaganda, made possible by their complete control of the channels of communication, but violence and the threat of violence were freely engaged in by these same authorities in an avowed effort to control *all* aspects of human thought in the directions *they* believed to be consonant with the best interests of the whole society. *Total* control was aimed at, and everything was supposed to be done in the interest of the *total*—the whole. Any interest not oriented toward the supposed best interest of the whole was considered to have no rightful standing. Such interests might or might not be permitted to exist, depending on a number of factors; but they could never, per se, lay claim to respect.

By contrast, our free societies were supposed by their apologists at that time to be states in which *no* such attempts at thought control were ever made and in which any and all meanings were supposed to be "thrown into the marketplace of ideas" to be "bought"—or passed up—by each individual as he chose, freely and on whatever grounds. The free society was therefore sometimes described as the "open" society—notably by Sir Karl Popper—as against the "closed" one espoused by the totalitarians. However, as apologists for the totalitarians rightfully pointed out, our free societies were by no means so open as this theory maintained. We advocates of free societies were therefore accused of hypocrisy. It was pointed out that in fact there were numberless limits upon the vaunted openness of our societies. Many traditions put severe restrictions upon men's freedoms. These restrictions, it was pointed out, were in many cases even enforced by the governments of our free societies.

Advocates of our free societies therefore took to apologizing for the existence of the traditional limits on our freedoms, that is, for our values and our morals. It became fashionable to call these values and morals regrettable—but temporarily unavoidable—imperfections marring our open societies and to hold that these societies, while not yet wholly open, would be someday (a gratuitous argument, used, of course, in almost identical terms by the apologists for the Stalin horrors). Many advocates of free societies therefore fell with a will to attacking any and all traditional restrictions and, in doing so, strengthened the skepticism and the nihilist thinking that, as we pointed out at the beginning of this work, were the seedbed from which the totalitarian movements had sprung. They failed to realize that a free society *rests upon a traditional framework* of a certain sort; and, in their mistaken notion that a free society is an open one, they threw out the baby with the bath water.

This is a very serious mistake. A *wholly* open society would be a wholly vacuous one—one which could never actually exist, since it could never have any reason for existing. What we must now do, therefore, is something not likely to prove popular with a great many contemporary intellectuals. We must show the need for a traditional framework in a free society and must try also to show what kind of traditional framework it must be if the society is to be free.

To make the task easier, and to see more clearly what is meant, we can take as an example a modern endeavor of great importance; this will then serve us as a paradigm for other intellectual and moral enterprises in our free society.

Let us take as our example the pursuit of modern science. This example is a telling one, since modern science was founded on a violent rejection of all authority. The revolt against authority was its battle cry throughout its formative period. It was sounded by almost every prominent early scientist and by their philosophical apologists— by Bacon and Descartes and Locke and Hume, among many others. It also was chosen by the founders of the Royal Society as their motto: *Nullius in Verba*. This battle cry, as they originally understood it, was meaningful, since in their day it was important for them to reject all external authorities, inasmuch as all the traditional authorities at that time were adversaries of the new science and had to be rejected. Once these opponents were defeated, however, the slogan remained and came to imply that the pursuit of science required the repudiation of *all* authority and of *all* tradition. It became very misleading at this point, since, as we shall see, the pursuit of science certainly does not and cannot repudiate all authority and tradition.[1]

The popular conception of science says that science is a collection of observable facts that anybody can verify for himself. We have seen that this is not true in the case of expert knowledge, like that needed in diagnosing a disease. Moreover, it is not true in the physical sciences. In the first place, for instance, a layman cannot possibly get hold of the equipment for testing a statement of fact in astronomy or in chemistry. Even supposing that he could somehow get the use of an observatory or a chemical laboratory, he would not know how to use the instruments he found there and might very possibly damage them beyond repair before he had ever made a single observation; and if he should succeed in carrying out an observation to check upon a statement of science and found a result that contradicted it, he could rightly assume that he had made a mistake, as students do in a laboratory when they are learning to use its equipment.

The acceptance of scientific statements by laymen is really based, not on their own observations, but on the authority that laymen acknowledge scientists to have in their special fields; and this is true to nearly the same extent of scientists using results of sciences other than their own: they do not feel called upon, or even competent, to test these results themselves. Scientists must rely heavily for their facts on the authority they acknowledge their fellow scientists to have.

The authority scientists thus have is even exercised in a personal manner in the control they exert over the channels through which contributions are submitted to the scientific community. Only offerings that are deemed sufficiently plausible are accepted for publication in scientific journals, and what is rejected will be ignored by science. Such decisions are based on scientists' fundamental convictions about the nature of things and about the method which is therefore likely to yield results of scientific merit. These beliefs and the art of scientific inquiry based on them are not codified into laws and regulations, nor are they applied in anything like a legalistic manner. They are, in the main, everything that is involved in the traditional pursuit of scientific inquiry, and they are to a great extent only tacitly applied in forming a judgment.

To illustrate our meaning, let us take as an example of a claim lacking plausibility to the point of being absurd a letter published some twenty-five years ago in the journal *Nature*. The author of the letter had observed that the average gestation period of different animals, ranging from rabbits to cows, was an integer multiple of the number π. The evidence he produced was ample, the agreement good. Yet the acceptance of this contribution by the journal was meant only as a joke. No amount of evidence would convince a modern biologist that gestation periods are equal to integer multiples of π. Our biologists' conception of the nature of things tells them that such a relationship is absurd, but it does not at all prescribe how to prove this. Another example—from physics—can be found in a paper by Lord Rayleigh, published in the *Proceedings* of the Royal Society in 1947. It described some fairly simple experiments which proved, in the author's opinion, that a hydrogen atom impinging on a metal wire could transmit to it energies ranging up to a hundred electron volts. Such an observation, if correct, would be far more revolutionary than the discovery of atomic fission by Otto Hahn in 1939. Yet, when this paper appeared, physicists were not impressed by it. They merely shrugged their shoulders. While they could find no fault with the experiment, they not only did not believe its results but did not even think it worthwhile to consider what was wrong with it, let alone check up on it. They just ignored it. About ten years later some experiments accidentally offered an explanation of Lord Rayleigh's findings. His results were apparently due to some hidden factors, of no great interest, which he could hardly have identified at the time. He should have ignored his observation, for he, as a physicist, ought to have known, as his colleagues did, that there must be something wrong with it.

It is true that the rejection of implausible claims has often proved mistaken; security against this danger could be assured, however, only at the cost of permitting journals to be swamped by nonsense. Their editors would have to suspend their principles of judgment and publish *everything* submitted.

These principles of judgment that are used to screen articles for publication are largely traditional, since they have been acquired by individual scientists from their mentors and from the literature on their subject, and they are, at bottom, for the most part only tacit understandings. Even when attempts are made to state them explicitly, what these explicit statements *mean* can be known only by scientists in the particular field involved. There is much that cannot be made explicit because it lies at the level of feelings about fitness and in working attitudes that betray an essentially imaginative grasp of how things in that field may be expected to work or to *be*.

There are some very general principles of scientific judgment that can be stated, of course; but their statement will itself prove our point. Much has been said, popularly, about the exactitude of acceptable scientific results. Certainly exactitude is a general principle that is applied in ascertaining scientific merit. But what counts for exactitude differs in different fields. Measured by any absolute scale, the exactitude of results is much less in some fields than in others. There are at least two other principles that also enter into scientific judgments; and when the results achieve a high rating in terms of these principles, a lower level of exactitude is regarded by scientists as acceptable.

These two principles are *systematic importance* and the *intrinsic interest of the subject matter*. Even in physics absolute exactitude is seldom found, but requirements for exactitude will be less if the results seem to fit beautifully into a system of theories that are already well accepted or if they can serve to further the systematic development of theories. In many cases there can be little intrinsic interest in the inanimate—and often inexperienceable—nature of the subject matters dealt with in physics. By contrast, such intrinsic interest may be quite high in the subject matters dealt with in zoology and botany, and results in these sciences will therefore be respected even when they are relatively inexact and contribute very little, by comparison with results in physics or chemistry, to the systematic development of the science.

Science is shaped by the delicate evaluations scientists make in each of these areas. They judge contributions in terms of these values of accuracy, systematic interest, and what would almost have to be called the *lay* interest in the subject matter; weighing these together, they make their judgments in terms of an optimum they and their peers have tacitly come to accept. The application of this optimum, like the application of an economic optimum, can result in a decision; but the process of its application cannot be specified, because the relative values of the various factors have no determinate existence before the judgment: they are assigned only *in the making of the judgment*. These judgments are tacit judgments, therefore, and always personal. But they are not whimsical. The personal taste in accordance with which they are made leans heavily upon a traditional taste acquired by the scientist in his acculturation to the community of scientists he has succeeded in joining.

If this view of the extent to which scientific judgment rests on tradition is correct, we may wonder how the conformity so enforced can allow for the appearance of any true originality. For it certainly does allow it: science presents a panorama of surprising developments. How can such surprises be produced on such dogmatic grounds?

We often hear of surprising confirmations of a theory. The discovery of America by Columbus was a surprising confirmation of the theory of the earth's sphericity; the discovery of electron diffraction was a surprising confirmation of De Broglie's wave theory of matter; the discoveries of genetics brought surprising confirmations of the Mendelian principles of heredity. We have here the paradigm of all progress in science: discoveries are made by pursuing possibilities suggested by existing knowledge.

This applies also to radically novel discoveries. All the material on which Max Planck founded his quantum theory in 1900 was open to inspection by all other physicists. He alone saw inscribed in it a new order that could transform the outlook of man. No other scientist had any inkling of this vision; it was more solitary even than Einstein's discoveries. Although many striking confirmations of it followed within a few years, so strange was Planck's idea that it took eleven years for quantum theory to gain final acceptance by leading physicists. Yet, in another thirty years, Planck's position in science was approaching that hitherto accorded only to Newton.

While science imposes an immense range of authoritative pronouncements, it not merely tolerates dissent in some particulars but grants its highest encouragement to such creative dissent. While the machinery of scientific institutions severely suppresses suggested contributions when they contradict the currently accepted view about the nature of things, the same scientific authorities pay their highest homage to ideas whose implications are destined sharply to modify these accepted views.

This apparent contradiction is resolved by the circumstance that all our knowledge of the external world actually rests on tacitly accepted metaphysical grounds. The sight of a solid object indicates to everyone that it has both another side and a hidden interior, which we could explore; the sight of another person indicates to us a set of unlimited hidden workings of his mind and body. Perception has this inexhaustible profundity, because what we perceive is tacitly understood by us to be an aspect of reality, and aspects of reality are tacitly believed to be clues to boundless undisclosed, and perhaps yet unthinkable, experiences. This is what the existing body of scientific thought means to the productive scientist, whether or not he will verbalize it in this way. He sees in it an aspect of reality which, as such, promises to be an inexhaustible source of new and fruitful problems. And his work bears this out: science continues to be fruitful because it offers an insight into the nature of reality.

This view of science merely recognizes what all scientists believe: that science offers us an aspect of reality and may therefore manifest its truth inexhaustibly and often surprisingly in the future. Only in this belief can the scientist conceive problems, pursue inquiries, and claim discoveries. This belief is the ground on which he teaches his students and exercises his authority over the public, and it is by transmitting this belief to succeeding generations that scientists grant their pupils independent grounds from which to start toward their own discoveries, possibly in opposition to their teachers. But we must also note that this reality is, of course, always imagined as structured in a certain way; and these current beliefs about the *fundamental* nature of reality are the working context in which any change, even any "novel" change, is judged. If its novelty threatens *these* current basic views, it will have a hard time getting a hearing, much less winning acceptance; yet it may do so in time if some scientists, noticing that it offers vistas of meaningful possibilities richer than their present fundamental views do, become converted to it.

The discovery of new facts may also change the interest scientists take in certain established facts, and intellectual standards themselves are subject to change. Interest in spectroscopy was sharply renewed by Bohr's theory of atomic structure, and the novelty of its appeal also wrought a change in the standards of scientific beauty. No single achievement has equaled Planck's discovery of quantum theory in its effect upon the transformation of the quality of intellectual satisfaction in mathematical physics. Such changes have been accompanied through the centuries by the belief that they offered a deeper understanding of reality, and only in this conviction can a scientist inaugurate novel standards that are not merely his own private predilections but are offered with responsible universal intent, i.e., with the belief that they will prove to be what others also can accept as true. With these beliefs in the back of his mind, a scientist can *both* teach his students to respect current values in science *and* encourage them to try one day to deepen these values in the light of their own insights.

As a practicing scientist for many years, I believe all this to be a true description of present scientific procedure. But a true description of present scientific procedure implies no justification of it. However, if one trusts, as I do, that the metaphysical beliefs of scientists necessarily assure discipline and foster originality in science, one must declare these beliefs to be true. And I do this. Yet this does not mean that I share *all* accepted beliefs of scientists about the nature of things. On the contrary, all my writings show that I dissent in many areas, particularly in psychology and sociology; but this dissent does not mar in any way my notions about the necessary dedication of scientists toward external truths. Because we understand one another to be so dedicated, we can respect, not try to stamp out, one another's differing views.

Unfortunately, these metaphysical beliefs are not explicitly professed today by scientists, let alone by the general public. Modern science arose by claiming to be grounded in experience and not on a metaphysics derived from first principles. Our assertion that science can have both discipline *and* originality only if it believes that the facts and values of science bear on a still unrevealed reality stands in opposition to the current philosophic conception of scientific knowledge. These current philosophic views, we trust, have been shown to be short of the true commitments that scientists actually have.

The freedom of scientists to make truly original contributions has thus been shown to rest on various traditional beliefs enforced by the community of scientists as a whole. These beliefs make for continuity in both discipline and innovation and bring them together.

We next have to show how this community of scientists actually governs itself in accord with these principles. Research is pursued by thousands of independent scientists all over the planet, each of whom really knows only a tiny part of science. How do the results of their inquiries, each conducted largely in ignorance of the others' work, sustain the systematic unity of science? And how can many thousands of scientists, each one of whom has a detailed knowledge of only a very small fraction of science, jointly impose equal standards on the whole range of vastly different sciences?

Today's system of science has grown from the system it was a generation ago through advances which originated independently at a great number of points where the old system offered a chance for progress. Scientists were looking out for such points, and each developed one. Each studied the work of others on various promising points and also considered how he could best make use of his own special gifts. Such a decentralized and free procedure of mutual adjustment through self-coordination achieves the greatest total progress possible in practice and best assures the systematic character of science at successive stages of its progress.

All institutions serving the advancement and dissemination of science rely on the supposition that a field of potential systematic progress exists, ready to be revealed by the independent initiative of individual scientists. They would soon lose their *raison d'être* and degenerate into mere special-interest factions if they ceased to rely on such a supposition. It is, for instance, in view of this belief that scientists are appointed for life to the pursuit of research and are granted permanent subsidies for this purpose. Many expensive buildings, pieces of equipment, journals, etc., are founded and maintained in this belief. It is the most general traditional belief which a novice accepts when he joins the scientific community.

This raises another, more intricate problem that such a community must solve. How can we confidently speak of science as a systematic body of knowledge and assume that the degree of reliability and intrinsic interest of each of its branches can be judged by the same standards of scientific merit? Can we possibly be assured that new contributions will be accepted in all areas by the same standards of plausibility and be rewarded by the same standards of accuracy and originality and interest? Unless contributions are accepted in different areas by substantially equal standards, a gross waste of resources might ensue. Can such a scandal be guarded against by transferring resources from areas where standards are currently lower to points at which they are higher?

It might seem impossible to compare the complex scientific value of marginal contributions over such different areas as, for example, astronomy and medicine. Yet this is in fact done or at least is reasonably approached in practice. It is done by applying a principle we can call the *principle of mutual control.* It consists, in the present case, of the simple fact that scientists keep watch over one another. Each scientist is both subject to criticism by all other scientists and encouraged by their appreciation of him. This is how the scientific opinion is formed which enforces scientific standards and regulates the distribution of professional opportunities. It is clear that only fellow scientists working in closely related fields are competent to exercise direct authority over one another; but their personal fields will form *chains of overlapping neighborhoods* extending over the entire range of science. It is enough, of course, that the standards of plausibility and worthwhileness be equal at every single point at which the sciences overlap to keep them equal over all. Even those in the most widely separated branches of science will then rely on one another's results and support one another against any laymen seriously challenging their authority.

Such mutual control produces a *mediated* consensus among scientists even when they cannot understand more than a vague outline of one another's subjects. This also applies, of course, to myself. All that I have said here about the workings of mutual adjustment and mutual authority is based on my own personal belief that the modes of intercourse I have observed in my part of science can be assumed to extend through all sciences.

Mutual control applies also to those newly joining the scientific community at any particular point of its vast domain. They start their inquiries by joining the interplay of mutual coordination and at the same time taking their own part in the existing system of mutual control, and they do so in the belief that its current standards are essentially true and common throughout science. They trust the traditions fostered by this system of mutual control without much experience of it and at the same time claim an independent position from which they may reinterpret and possibly revolutionize this tradition.

The scientific community incorporates a hierarchy, to be sure, but this feature does not alter the fact that the authority of scientific opinion is exercised by the mutual control of independent scientists, far beyond the scope of any one of them. Scientists with worldwide reputations, editors of scientific journals, and senior members of academic faculties have indeed far more influence than the average scientist or the newcomer. But such a hierarchy is itself established mainly by the general respect freely tendered to their opinions by members of the scientific community.

We have seen that the scientist can conceive problems and pursue his investigation only by believing in a hidden reality on which science bears. Now that we have shown, further, how scientific originality springs from scientific tradition and at the same time supersedes it, we can show how this process establishes the sense of personal responsibility which sustains the scientist's search.

There are two possible ways of viewing the progress made by the front line of scientific discoveries as it advances over a period of time. We may look back upon such progress as the growth of thought in the minds of gifted people along the determined or "caused" pathways of science. The frequent occurrence of simultaneous discoveries may appear to support this image. Even major discoveries, which fundamentally refashion our conception of nature, can be made simultaneously by a number of scientists at different places. Quantum mechanics was discovered in 1925 by three authors so independent of one another that they were thought at the time to have given mutually incompatible solutions to the problem. Thus seen, the growth of new ideas appears altogether predetermined. The minds of those making discoveries seem merely to offer a suitable soil for the proliferation of new ideas.

Yet, looking *forward*, before the event, the act of discovery appears personal and indeterminate. It starts with the solitary intimations of a problem, of bits and pieces here and there which seem to offer clues to something hidden. They look like fragments of a yet unknown coherent whole. This tentative vision must turn into a personal obsession; for a problem that does not worry us is no problem. There is no drive to it. It does not exist. This obsession, which spurs and guides us, is about something that no one can describe: its content is indefinable, indeterminate, strictly personal. Indeed, the process by which this unknown thing will be brought to light will be acknowledged as a discovery precisely because it could not have been achieved by persistent application of explicit rules to given facts. The true discoverer will be acclaimed for the daring feat of his imagination, which crossed uncharted seas of possible thought.

Thus the backward-looking picture, of human brains as passive soil for the proliferation of thought, proves false. Nonetheless, it does represent an aspect of the pursuit of science. Scientific progress seen after the event may show the possibilities that were previously hidden and dimly anticipated in a problem. This does explain how different scientists may independently feel intimations of a particular potentiality, often sighting it by different clues and possibly discovering it in different terms.

There is a widespread opinion, conflicting with what we have just said, that holds that scientists hit on discoveries merely by trying anything that happens to cross their minds. This opinion results from a failure to recognize man's capacity for anticipating the approach of hidden truth. The scientist's surmises or hunches are the spurs and pointers of his search. Since they involve high stakes, they are as hazardous as their prospects are fascinating. The time and money, the prestige and self-confidence, gambled away in disappointing guesses will soon exhaust a scientist's courage and standing. His gropings are weighty decisions. They are not random flights from the top of his head in just any direction.

Choices made in the course of scientific inquiry are therefore responsible choices made by the scientist, but the object of his pursuit is not of his making. His acts stand under the judgment of the hidden reality he seeks to uncover. His vision of the problem, his obsession with it, and his final leap to discovery are filled from beginning to end with an obligation to an objective properly called "external." His acts are intensely personal acts, yet there is no self-will in them. Originality is guided at every stage by a sense of responsibility for advancing the growth of truth in men's minds. Its freedom lies in this perfect service.

Many writers have observed, since John Dewey advanced the idea at the close of the last century, that, to some degree, we shape all knowledge by the way we know it. Stated in this bald way, knowledge would appear to be subject to the whims of the observer. But the pursuit of science has shown us how, even in the shaping of his own anticipations, the knower is controlled by impersonal requirements. His acts are personal judgments exercised responsibly with a view to a reality with which he is seeking to establish contact. This holds for all seeking and finding of "external" truth, i.e., the attainment of "self-centered," as against "self-giving," meaning.

This is the only positive justification there is for accepting science as true. Attempts have been made to compensate for this apparent deficiency by reducing the claims of science from truth to probability. The uncertainty and transiency of science have been emphasized and exaggerated for this purpose. Yet all this is beside the point. The affirmation of a *probable* statement includes a judgment no less personal and no less directed by universal intent than an affirmation of its *certainty* would include. Any conclusion, whether given as a surmise or claimed as a certainty, represents a commitment of the person who arrives at it. No one can utter more than a responsible commitment of his own, and this completely fulfills his responsibility for finding the truth and for telling it. Whether or not it *is* the truth can be hazarded only by another, equally responsible commitment. There is no explicit or automatic way to avoid this necessity.

A scientist, having relied throughout his inquiry on the presence of something real, hidden out there, can rely only on that external presence also for claiming the validity of the result that satisfies his quest. As he accepted the discipline which the external pole of his endeavor imposed on him throughout his inquiry, he expects that others—if similarly equipped—will also recognize and accept the discipline of the presence that guided him. By his own command, which bound him to the quest of reality, he will claim that his results are universally valid. Such is the universal intent of a scientific discovery.

We are not holding that he has thereby *established* universality, but only that he has exhibited a universal *intent*, for a scientist cannot know whether his claims will be accepted. They may turn out to be false, or, even though actually true, they may fail to carry conviction. He may even suspect all along that his conclusions will prove unacceptable. In any case, their *acceptance* will not guarantee him their truth. "Acceptance" is not equivalent to "truth." To claim validity for a statement merely declares that it *ought* to be accepted by everyone, because everyone *ought* to be able to see it—as Galileo supposed that the churchmen of his day *ought* to be able to see what he saw if they looked through his telescope. The affirmation of scientific truth has an obligatory character; in this it is like all other valuations that are declared universal by our own *respect* for them.

We have spoken of the excitement of problems, of an obsession with hunches and visions that are indispensable spurs and pointers to discovery. But science is popularly supposed to be dispassionate. There is indeed an idealization of this supposition, current today, which deems the scientist not only indifferent to the outcome of his surmises but actually seeking their refutation.[2] This is not only contrary to experience; it is logical nonsense as well. The surmises of a working scientist are born of his imagination in the quest for a genuine discovery. Such an effort *risks* defeat by submitting these surmises to rigorous tests, but never *seeks* it. It is in fact his craving for success that makes the scientist take the risk of failure. There is no other way he can gain such success. As we have seen with respect to all of life, the category of success cannot exist without that of possible failure. Courts of law employ two separate lawyers to argue opposite pleas, because it is only by a passionate commitment to a particular view that the imagination can discover the evidence that supports it. A scientist's passionate commitment to a position does not first arise when he has failed a certain number of times to falsify that position.

The creative thrust of the imagination in science is fed by various

sources. The beauty of the anticipated discovery and the excitement of
its solitary achievement contribute to it in the first place. But the
scientist also seeks professional success; and, if scientific opinion
rewards merit rightly, ambition too will serve as a true spur to
discovery. Here we see a nonmoral principle, ambition, harnessed for a
moral end. The pursuit of science, as we have seen, is part of the moral
sphere of obligation, which we now can see must be a higher level of
the cultural realm (with principles of its own) resting upon a lower
level of the cultural realm—an essentially nonmoral level, the level of
ambition, the pursuit of power and profit.

We have now, in the instance of scientific inquiry, seen how a kind
of moral association of persons, through the exercise of mutual
authority, welds tradition and freedom together in a pursuit of the
truth and how actions of persons in this association are rendered
responsibly universal in intent by a common belief in the existence of a
reality, further and further aspects of which may be uncovered by
these persons through their own originative actions.

The pursuit of science, as we indicated earlier, can serve as a
paradigm for other free associations of persons dedicated to other ends
that are, like truth, conceived to be of intrinsic worth—i.e., ends that
are considered in some sense to be worthy of *respect*. Freedom can be
demanded by the dedicated individuals in these associations on the
grounds of that to which they are dedicated. It is only the pursuit of
truth, in the case of scientific associations, that can justify an
individual scientist's right to his freedom of inquiry and to the
publication of his findings.

This is a right he can demand from his colleagues on these grounds.
But on what grounds can he demand it from his fellow citizens and
from his government? It should be clear that he has no other basis
upon which to demand it from them, either, than a respect for truth
of the sort that is honored by his own community of scientists. He will
not have the freedom to pursue scientific inquiry, therefore, unless the
public also has respect for the ideal of scientific truth and *trusts* those
who are accepted as scientists to be engaged honestly in its pursuit.

We can see, therefore, that a free society is not simply an "open"
one, a society in which anything goes. It is a society in which men,
being engaged in various activities whose ends are considered worthy
of respect, are allowed the freedom to pursue these ends. A free society

is therefore one whose citizens in the main are committed to—dedicated to—various ideal ends (such as truth) and therefore one that is able to respect the free activities of its citizens in pursuit of such ends. It cannot be a free society by being open on matters such as these, that is, by being neutral with respect to truth and falsehood, justice and injustice, honesty and fraud.

We began this chapter by inquiring into the working nature of a scientific community in order to see if it could shed some light on the nature of a free society. This inquiry has now shown us how an association can be bound traditionally to certain standards and values and yet be free—both in the sense of being innovative and in the sense of being self-governing or autonomous. It has also shown us that a free association of this kind can exist only within a society operating very much along the lines it does and dedicated to upholding the same sort of ideals that it upholds.

Let us now see, in the last chapter, how such a free scientific community and the notion of mutual authority that governs it may be useful as paradigms for other free associations in a free society and, finally, for a free society itself.

THE FREE SOCIETY

THERE ARE OTHER ASSOCIATIONS OF PERSONS THAT REQUIRE FOR their existence the same sort of free activity that science does. All areas of scholarship operate along lines similar to those of science. In general, persons engaged in academic activities can demand freedom—to inquire and to teach—on the same grounds that scientists do, i.e., the pursuit of the sort of truth appropriate to each of these activities severally. The pursuit of all the arts and crafts likewise requires this kind of freedom. So do the profession of religion and the pursuit of justice under law, i.e., the administration of law. All these areas of free interaction operate within a tradition disciplining them but also making room for innovations in the sorts of mutual adjustments and criticisms individuals make *vis-à-vis* one another's activities.

Since we cannot explore all these activities in detail, let us take one that is dedicated to an end that everyone agrees is important—the administration of law, looking to justice.

A judge sitting in court and pondering a case refers consciously to many precedents—and perhaps unconsciously to many more. Numberless other judges have sat where he is sitting and have decided according to statute, precedent, equity, and convenience, just as he now sits and must decide. His mind is in constant contact with their minds as he analyzes various aspects of the case. Moreover, beyond the purely legal references he is making, he senses the entire contemporary trend of opinion, the social medium as a whole. When he has determined what bearing his knowledge of all these matters has upon the case and has responded to them in the light of his professional conscience, his declared decision will carry his conviction and will receive respect from his fellow members of the bar.

His decision will then, in its turn, affect public opinion and the decisions of future judges in future cases. The operation of law thus constitutes a sequence of adjustments among successive judges, guided by a parallel interaction between judges and the general public. What results is an orderly growth of the law through the application and reinterpretation of the same fundamental rules and their expansion into a system of increasing scope and consistency.

There are many important differences between the way individual scientists work within the community of scientists and the way judges work within the community of judges and lawyers and the general public, yet the two are similar in this: that when a scientist wrestles with a problem and accepts as his premise a great mass of previously established knowledge and submits to the guidance of scientific standards, while also taking into account the whole trend of current scientific opinion, he resembles a judge referring to precedent and statute and interpreting them in the light of contemporary thought.

The method of mutual adjustment common to both judges and scientists is clearly a process of *consultation*. There is, of course, another method of mutual adjustment, one used in another kind of associative activity where such adjustment is essential—an activity we have not mentioned as yet. This is business activity. Businessmen engage in mutual adjustment largely through *competition*, in which each is guided primarily by striving for individual advantage. The freedom each participant needs if he is to operate in modern economic activity is also justified by reference to the ends to be fulfilled, but there is a difference between these ends and the others we have been speaking of. The end of economic activity is to provide individuals with the material goods and services they want to use and enjoy individually. Economic ends are thus not spiritual or ideal ends like truth, justice, beauty—ends that can be shared by people (in fact, that can hardly exist at all unless they *are* shared) and are not used up in their enjoyment. The production of a splendid work of art or piece of music, the growth of a more coherent and consistent body of law, the development of a grand new theory in physics, stand as spiritual or ideal achievements in which we all share but which none of us consumes. But an automobile running off an assembly line or a bushel of wheat coming off a farm is nothing at all unless some particular individual eventually consumes it.

The freedom essential to the efficient and economical production, distribution, and consumption of economic goods and services cannot be argued for on moral grounds, therefore, since this freedom is not essential to the creation of that which is fraught with moral meaning and obligation, such as truths or justice or something that enhances or raises the meaning level—the significance of lived lives—as art and religion do. These latter ends are ends that, once grasped as meaningful, demand our responsibility to them—generate obligations to them; we therefore say that they are of *inherent* worth, are something we believe *ought* to be respected. Thus we can go on to say that the *freedom* to pursue these ends ought also to be respected.

We can say that the economic freedom to make mutual adjustments that are mutually acceptable ought to be respected also. But the "ought" here is conditional, not categorical. It depends upon whether or not we have an industrial mode of production requiring so many continual adjustments that central planning is an impossibility. The "ought" in this case is something like the "ought" involved in a hypothetical imperative (to use Kant's terms). It is like saying, for example, that, if one has a gas stove, one must connect it to a source of gas and light it before one can cook one's food. There is, of course, no moral obligation to connect or to light the stove—or even to cook one's food. By contrast, the other ideal ends, those which demand freedom for their accomplishment—and so, also, the freedom that is so demanded—would be more similar to Kant's categorical imperative. These command unconditionally—when they command at all.

There are myriads of particular problems, knotty difficulties, in maintaining these enclaves of free activity in a society. It is not our business here to get into detailed suggestions for their solution. As a matter of fact, such detailed suggestions, made in a book like this, would be idle. In a free society, these problems can be solved only by mutual political adjustment as time goes on. They are not technical philosophic problems that some academic thinker can work out solutions to in a library or while crossing the Quad.

What is important about our discussion concerning these enclaves of freedom is that we bring out the basic sense in which the institutions of a free society differ from those of its opposite. We must also bring out the way this difference is connected to what we as a people are passionately committed to.

We have been speaking of freedom in this chapter as though it were something desirable. But "freedom" is an ambiguous term, and in some of its meanings it has sometimes been severely criticized.

One very basic meaning of "freedom" seems to be "absence of

external restraint." The rational limits to freedom, understood in this fashion, are set by the condition that our exercise of a freedom must not interfere with other people's right to the same freedom. These are reasonable limits, because this sort of freedom cannot possibly exist in a social situation (i.e., be accepted as "right" by everyone in a society) without tacking these limits onto it. This sounds very simple, and certain examples can be adduced that are easily understood and acceptable to everyone. Take, for instance, the freedom to go to sleep or to watch TV. I should have that freedom, this principle of limitation says, as long as my exercise of it does not interfere with my neighbor's freedom to choose between the same alternatives. We have inherited this principle governing the use of individualistic freedom from the great Utilitarians of our past. They linked it to the idea that the pursuit of a good society is the same as the pursuit of the greatest happiness of the greatest number and that freedom in the sense we are discussing it here is a necessary condition for the effective existence of this pursuit.

At its base, however, this is an individualistic, self-assertive conception of freedom and, because this is what it is at bottom, it unfortunately can be used—and has been used—to justify all kinds of socially objectionable and even destructive behavior. The worst kinds of exploitation—of the poor, of children, of women, even the keeping of slaves—have been practiced in its name. It also has served as the ground for the Romantic movement, exalting the unique, lawless individual, and for those nations striving for "greatness" at any price.

There is another meaning for freedom that is almost the exact opposite of the one we have just been discussing. It regards "freedom" as "liberation from personal ends by submission to impersonal obligations." When Martin Luther faced the Diet of Worms and declared, "Here I stand and cannot do otherwise," he was not asserting his lack of freedom to do otherwise. He was maintaining that his acknowledgment of a moral demand gave him a freedom from the pursuit of merely personal ends (such as the protection of his own life) as well as, in this case, a freedom from having to obey the authorities in religious matters. This is, of course, a form of liberation, although it is quite different from the self-assertive, individualistic one of the Utilitarians; it is indeed, from their point of view, even foolish.

But such a meaning for freedom can become very much like a theory of totalitarianism. It becomes exactly that by the mere addition of the notion (not in Luther's intentions, of course) that the state is the supreme guardian of the public good. Then the dangerous paradox mouthed by all the totalitarians of our century follows: the individual is made free by surrendering completely to the state.

The preservation of freedom in the lives of men is thus gravely imperiled by both these conceptions of freedom; for even if men do not go to the extreme of either anarchy or totalitarianism under the thrust of these meanings, they may well feel, on the one hand, that the individualistic theory of freedom is simply selfish. At least it is uninspiring. Certainly the young men of Europe who were inspired by the totalitarian view of freedom to march to their deaths in World War II had found the individualistic view of freedom to be uninspiring. On the other hand, the theory of freedom as self-surrender to impersonal obligations does not seem to accord with our sympathy for the individual's pursuit of his own happiness in his own personal manner.

If we reflect now upon the structure that freedom was seen to have in the paradigms provided by the scientific community and the legal community, we can see how these existing communities do, in fact, weave into a working whole the two aspects of freedom we have just discussed. The freedom of a scientist or a judge is not one of simple self-assertion. It is a freedom to pursue certain obligations and to share in a system of mutual authority. Nevertheless, this freedom implies an absence of external restraint because it also entails a right to make personal judgments (often quite innovative)—judgments that bring whole segments of our person into them.

Let us now also try to use this paradigm of a scientific community to enlighten ourselves on what sort of structural principles a free society as a whole may need.

By a simple and obvious analogy, a free society must exist within the context of a tradition that provides a framework within which members of the society may make free contributions to the tasks involved in the society. The freedom of mere self-assertion can lead only to disintegration of our standards and institutions. It may from time to time result in an equilibration of social forces—interests—that mutually tone one another down to the point where they all can live together in some sort of working balance. This is something like what Madison hoped would be the case in the large society he expected the United States to become.[1] However, no one who holds the view that freedom is mere self-assertion will be devoted to maintaining such a balance; he will rather be devoted to upsetting it, whenever possible, in order to achieve more of his own interests. Thus any balance that happened to exist at any time would always be threatened and almost certainly would, from time to time, be completely destroyed. As Adam Smith foresaw, the chief danger to the optimal balance that could be achieved under institutions of free trade would come from the manufacturers, for none of them would have any interest in maintaining a free system of competition; their interest would rather lie in securing monopolies in order to control their markets. The adjustments that could and should be forced into existence by a free market would then not occur—in their case. Since all industrial producers would be doing the same thing, none of them would be doing what a free market would require. Smith thought, therefore, that only a tremendous effort to educate farmers and workers in their self-interest could ever bring a free market into existence or preserve an existing one.[2] Actually, he seemed to have little hope of its success.

What needs to come into the picture of a viable free society is a traditional devotion to the spiritual objectives, such as truth, justice, and beauty—those that require for their pursuit free, self-determinative communities: of scientists, scholars, lawyers, and judges, artists of all sorts, and churchmen. For without a general public devotion to these spiritual objectives, free, self-determinative communities could not long continue to exist. The public (or public officials) would most certainly decide at some point to try to control these pursuits in the interest of the "general welfare." Of course, a public which succumbed to this temptation would soon have little or nothing to use for increasing the general welfare, for it would have inhibited, if not annihilated, *real* inquiry, *real* spiritual or moral insight, *real* justice, and *real* art. What it would have left, in the caricatures of these activities, would be powerless because meaningless. However, a too explicit and "official" devotion to these ideals (defining them too explicitly and setting up public agencies to promote them) would also destroy them, because it would destroy the freedom of people in these fields to make innovative mutual adjustments relevant to their pursuit. Such a society would to some degree become ossified into a rigid set of meaningless (perhaps mainly verbalized) objectives. Thought police would find an important niche in such a society as much as in the one in which the pursuit of ideal ends is subordinated to something called the "public good." What would have been forgotten is that such ideals as scientific truth, justice under the law, and good art cannot be given concrete definitions. What these really are, *in concreto*, is simply what all members of each relevant group are striving together to delineate. Truth, for instance, is given specific form only as the community of scientists is free to work out what its form is—and this task is never finished. The same thing is true of justice in the practical development of legal systems and of art in the continuing work of artists.

These enclaves of freedom—science, the law, art, and so forth—will have to consist of autonomous circles of men, free from public control to work out their problems through mutual adjustment and authority; in other words, they must be little republics of their own. The public must respect them sufficiently to refrain from trying to direct them in any way toward something called the "public good." Even the economy, since it *is* an industrialized economy, must be allowed its freedom to operate, as we have seen, by mutual adjustment of its participating parts through the mechanisms of markets and pricing and profits. Interference with this system of mutual adjustments can and must be made, of course, from time to time for various purposes of great importance to the public or for the preservation of the system of mutual adjustments itself; but attempts to supplant it altogether by central planning would simply bring an industrial society to a halt, as well as place power to control all activities of groups and persons in the hands of public officials, since all working capital (resources) would be controlled and distributed according to their judgments.

It would seem to be clear, under the concept of the free society that we have been outlining, that many of the affairs of the society would be managed through the development of various *spontaneous* orders—ordered wholes that develop freely by means of mutual adjustments, rather than *corporate* orders, ordered wholes given their shape from the top down, i.e., centrally controlled. It is our contention that a system that develops from the bottom up, through free interaction of its parts upon one another (subject only to a free, common dedication of its participants to the value of certain standards, principles, and ideal ends), is the only social system that can meaningfully be called free. The alternative is to control social affairs centrally, from the top down, and so establish a corporate order—which is the essence of totalitariansim.

Such a system of spontaneous order(s) is, however, open to several serious objections. Let us see what these are and attempt to answer them. It can be claimed that, under a system of spontaneous order: (1) The public good is surrendered to the personal decisions and motives of individuals. (2) Society is submitted to the rule of a privileged oligarchy. (3) Society is allowed to drift in a direction willed by no one.

As we have seen, under this system of spontaneous order, individuals—whether producers or consumers in the economy—are engaged frankly in the competitive pursuit of personal gain. Scientists, judges, scholars, clergymen, et al., are guided by systems of thought to promote the growth, application, or dissemination of that to which they are dedicated. Their actions are thus determined by their own professional interests, which do not aim specifically at promoting the general welfare of the society. The businessman must seek profits, the judge must find and apply the law, the scientist must pursue discovery, for that is what makes them a businessman, a judge, or a scientist. Each is really ignorant of how his action will affect the public good, nor could he allow himself to be deflected from his professional duties by such knowledge if he did possess it. We justly blame those judges in Nazi Germany who allowed their best legal judgment to be deflected by what they took to be the national interest at the time: the rebirth of a German national spirit and the presumed contribution made to this by the "temporary" designation of scapegoats for Germany's weak and demoralized condition.

As to the second objection, that society is submitted to the rule of a privileged oligarchy, it must be admitted that the system of spontaneous order involves the exercise by various elites of very considerable power to affect the public. Under a competitive market economy—generally called "capitalism"[3]—businessmen handle the major part of a people's wealth and direct the day-to-day activities of the people who are engaged in producing it—the workers. The social interests entrusted to an independent judiciary and to the free pursuit of science are no less momentous. Indeed, the mental activities cultivated by various members of the writing profession—poets, journalists, philosophers, novelists, preachers, historians, economists—are perhaps the most decisive in shaping public affairs and sealing the fate of society.

The activities of such persons as these in the systems of free, spontaneous order which we have just outlined may well give them the appearance of an oligarchic regime engaged in usurping the public power and short-circuiting the possibility of free, democratic public control. When we add to this the personal advantages these people possess in virtue of their positions, their prerogatives must seem even more invidious. The inheritance of property, plus the enhanced opportunities offered to the children of more highly placed parents, tends to make their position of power and privilege hereditary—the possession of a restricted class of families known, in Marxian terms, as the bourgeoisie. In a system of spontaneous order the public interest escapes control by the state but does so by being submitted to the control of what would surely seem to be an irresponsible bourgeois oligarchy.[4]

However, the members of this oligarchy, although they draw considerable benefit from the exercise of their functions, do not exercise control in anything like a planned or deliberate fashion. They literally do not know where they are going and cannot control the direction in which their separately achieved spontaneous orders are moving the whole society. This direction is not one specifically willed by them or by anyone. It *occurs*. In this system, it must be said, society is moving toward unknown destinations.

Under this way of ordering things, the state, though it must pay the judges and subsidize scientific research, cannot make the judges come to certain decisions that the state thinks are in the public interest, nor can a state that really wishes to promote scientific inquiry designate what inquiries should be undertaken. To the extent that states do try to direct such inquiries, they either inhibit progress in science or they make it necessary for the recipients of such funds to become liars and cheats. If states should be successful in directing inquiries in certain directions, they would still, of course, be unable to predict or control what discoveries would be made. The social consequences of any discovery are also unpredictable. They may turn out to be very beneficial or very harmful; probably they will be both to some degree.

What all this seems to mean is embodied in the third objection: that under this system we are adrift. But this can hardly be a criticism of the system, for the plain fact seems to be that man necessarily *is* adrift. The future *is* beyond our control, since it is beyond our comprehension. How should we be able to predict discoveries—which *are* discoveries precisely because no one could have deduced (pre-dicted) them from what was already known? But discoveries, as we know, can change the whole direction of human living. If a library of the year 3000 A.D. should come into our hands today, we would have every reason to believe that we would be unable to understand its contents. The future will be the meanings that men then will have achieved which men now have not yet achieved. We cannot *plan* what these shall be. We would have to know them in order to plan them. At any rate, we can control only if we have objectives. Therefore, we could strive to control the future only in terms of our own *present* objectives, since these are all we have; but when the future comes, men will deal with it in terms of the objectives they have, by then, moved forward to—not the ones we have today *planned* that men by then will have moved forward to! We cannot plan now for something that no one may wish to plan for then.

The system of mutual adjustments achieved at any one time is therefore good only for that moment. The system of mutual adjustments must be in continual flux as it moves to higher and higher levels of meaning; each advance entails a continual reassessment of the system as new factors and new meanings make their appearance. The achievement of any meaning means that the achievement of a further meaning must now be undertaken; for the existence of this new meaning is an *addition* to the meanings that were understood to be there when it was born, and our view now must be broadened to include the coming-into-being of this new meaning as well.

The progress of this system of continual reflexive adjustments cannot, of course, be known before it is known and therefore cannot (logically cannot) be planned for. But this does indeed seem to be the ontological situation of man in the world; if it is not so for all time, as it certainly seems to be, then at least it is his situation as of now.

It is this logic of man's situation that prevents him from controlling the drift of history. He can either embrace a social philosophy that fails to see this logic and tries to control the future (futilely, of course), or he can admit his powerlessness with respect to the future—his limits as a mere human—and embrace a social philosophy that upholds those institutions that do in fact leave the future unplanned—and free to make its own mutual adjustments.

This same logic can also justify the oligarchic system we have just seen to be involved in a free society, for it shows that the social tasks we have been discussing—science, law, art, etc.—can be achieved only by independent mutual adjustments. They are, in a word, *polycentric* tasks. They can reach a solution only if they are worked at from many centers, free to interact continually with one another in the formation of a system of mutual adjustments. This means that these tasks can be accomplished only through a spontaneous order, not through deliberate efforts to order them corporately. In terms of a very simple example, to secure that order among potatoes in a sack that will accommodate the greatest possible number of potatoes, one cannot sit down and plan (i.e., cannot corporately order) how each potato shall lie relative to every other potato and then direct them one by one into these positions. One need only give the bag a kick or two, and each potato is put in motion and mutually "settles in" with every other potato, achieving spontaneously an order of close proximity. By analogy, what "kicks" or prods scientists, judges, etc., into these systems of spontaneous social order are their private motives for entering initially into the "motion" involved in these various enterprises. But the motive power that will move them into the "best" relations to one another (the spontaneous order actually achieved) is the body of *standard* incentives that defines the nature of their activities. These incentives are one and the same as their professional duties. So, regardless of the *private* motives that move a person to be, for example, a judge—ambition for status, for power, for respect, for money, or whatever—he is *not* a judge unless he performs according to the standard incentives of that profession, and not according to his own private motives for entering it. These standard incentives are: to find the relevant law and the relevant facts and to make a decision that either follows the precedents or creates a new precedent on grounds that his colleagues can find (or at least *ought* to find) reasonable.

The foregoing observations show us that the individuals in these elites will operate on two levels: one is the lower level of ambition—of power and profit; the other is a higher level of moral obligation. The first can be seen to be lower in three ways. First, it is not as rich in principles as the moral level resting upon it; this is analogous to the way that, as we noted above, arithmetic is not as rich as calculus, which in a sense rests upon it. Second, when one is operating at the moral level—is "carried away" by it—one sees that the moral level is "higher," i.e., that it is more worthy of respect than the level of pure ambition, although one continues to *understand* the nature and the existence of the level of ambition and, indeed, does not need to deny it attention. By contrast, if one could imagine an operation carried on strictly at the level of ambition, one could do so only by banishing the moral meanings altogether from one's vision. Third, the moral level can exist only if it rests on the lower. The lower provides the private proddings or "kicks" that individuals who are to act morally in various positions must have in order to act at all. But these proddings also put limits to moral attainment. Just as our bodies make possible the intellectual systems within which we dwell (we cannot think without a brain), so also they set limits to our attainments in these systems. This is why we can never be perfect at anything—whether it is science, art, morality, or religion. The notion of perfection in any pursuit is an imaginative projection of what the full and unlimited operation of the principles governing these higher mental levels would look like. But the fact is that the operation of these higher mental levels is rooted in the physiological levels of our bodies. So our minds, in their actual operation, can never achieve independence from our bodies—just as the operating (boundary) principles of a machine are not, *in the operation of the machine*, independent of the physical laws governing the parts of the machine. It is in fact these physical laws governing the parts that make it possible for the operational principles to operate. But these physical laws also place limits on the operation. An automobile can be moved and stopped in a controlled and purposeful fashion. It can be improved by the development of better and better boundary or operating principles that structure its parts, so that it can move faster and stop more quickly. One might say that its "private" motivating powers (those of its physical parts, given intelligible structure by the laws of physics and chemistry) have been organized in a better way to accomplish these tasks. But these very laws also dictate limits to these accomplishments. Because of the way matter "works," an auto cannot be made to move at infinite velocity or to stop instantly, although this behavior would be the fullest imaginable projection of the intentions of the use of the machine, i.e., of its boundary principles, taken in abstraction from the "private" action of the parts that these boundary conditions have organized.

In a way perfectly analogous to all these examples, the moral sphere of man's life is made possible—but also is limited—by the systems of power and profit and restricted parochial interests upon which his moral sphere rests.

We have noted that persons who assume the responsibilities of various social tasks must have some private incentives to induce them to do so and that their moral action therefore rests upon a basis of personal power and profit. But the moral actions of these elites rest upon organizations of power and profit in a social sense as well. The various elites manning the cultural projects that raise mankind to a level above that of the other animals (in terms of the achievement of meanings that are far beyond the capacity of these other animals) *can* do this only if they have the *power* to operate freely in their own spheres. It may be true that the public acquiesces in their having this power only if they make responsible use of it; still, the public will destroy the several powers of these elites—and therefore their accomplishments—if it tries to closely supervise their use of these powers. A judge, for instance, either does or does not have the right to rule in a case; therefore, his *power* to do so either is or is not acknowledged by the public. If his power is not acknowledged, he does not have the *right* to rule in a case; and if there is no one belonging to the legal community who has this power or right, then there are no real judges—only public officials ruling by decree—and there is no rule of law in that community.

Since a judge must, therefore, have power to make rulings—power he is acknowledged to have by others who have political power—he must realistically ask, from time to time, a somewhat nonlegal question: Who will continue to support him—*and his judgments*—since he cannot enforce them himself? As Mr. Dooley said, "The Supreme Court follows the elictions." One eye cast in the direction of political considerations is the price of a judge's continuing to sit and to have the opportunity to dispense such justice as the public will bear. Yet this point does *not* "reduce" the principles of justice to the principles of power and profit. Even the Marxists have held that ideologies, once called into existence, have a life of their own—have their own principles; however, the only reality that the Marxists accord to these principles is that they reflect the interests of the ruling class in maintaining its position. We, on the contrary, claim only that the interests of the ruling class—i.e., those in the society who do possess political power at the moment—may result in *limiting* the extent to which justice can be actualized. These interests do not *determine* what shall be thought at any time to be justice ideally. *Judges* determine this through the long line of mutual adjustments in legal thinking they have been engaging in up to the present day.

We began this chapter by asking what kind of society will honor and respect the meaning achievements of man. We have discovered that it is a society made up of a number of associations of free, self-governing persons, participating by mutual adjustment in the pursuit of various ends, some of them thought to be of such intrinsic worth that they create obligations to themselves, i.e., they engender respect for themselves.

Moreover, these ends—and these participants—have to have the respect of the general public before the public will allow them to play these kinds of "games," even though no clearly discernible public good can be seen to come from them. Where commitments to ends of these sorts exist in a society, there a set of free associations may exist. Where they do not exist, there free associations *cannot* exist, or at least cannot exist for long.

Now we must ask how, in a free society, the public in general functions with respect to its government. This must also be by mutual adjustment and mutual authority, of course, for otherwise the citizens would simply be under the authority of others. The form this mutual adjustment takes is persuasion. This form is also used in various other mutually adjusting systems of spontaneous order, such as the efforts of scientists to get their colleagues to acknowledge their claims to an important discovery.

"Persuasion," in this case, must not be confused with a behavioral "change of mind." A mere change of mind might be accomplished by means of subliminal gimmicks, of lies, fraudulent claims, or, in general, behavioral "brainwashing." But this would not be persuasion as practiced by scientists. Their sort of persuasion is of an altogether moral kind, since it implies an "open-eyed" change of mind, one made without pressure, threats, or psychological tricks—in other words, one made in face of the facts and representing what the person *ought* to think when he has all the facts at his disposal and when no one is trying to conceal anything from him.

This, as we said, is the ideal of persuasion in scientific argument. It would be the ideal also in political persuasion among and between citizens, except that in political situations there are extenuating circumstances for departures from this ideal. Truth is the highest object or intention in science—the greatest standard motivation. In political persuasion the highest standard of motivation, perhaps, is what is right or fair or just in a distributive sense; but this can never be separated from the motive of interest, nor this, in turn, from the motive of power. Many interests exist in any society. None of them can be furthered without power. The members of each interest group will therefore think it is "fair" or "just" for them to have the power to further their own particular interests, and so their own need to enhance or protect their power will necessarily enter into what they think is "fair" or "just." And what can be agreed upon as fair or just *by the participants* in a society must become what settles political problems justly there. It would be difficult to know what it would mean to establish scientifically what is politically "just," that is, establish it by pursuing the ideal *truth* about it. Justice is always a practical or political problem, not a theoretical one. Even the communities of persons pursuing ideal ends (scientific, artistic, etc.) do not engage in a scientific or aesthetic inquiry into the truth about what their share should be when they compete for public funds.

If there could be a *truth* about which particular interests *ought* to have power, or ought to have *more* power than the others, then persuasion in political matters could have truth also as its highest standard incentive, as in science. If this were so, the public interest could be known by the elite whose particular aptitude and discipline it was to inquire into these matters, and only *these* persons should then *rule*. Persuasion of the scientific sort would then operate, but only in the mutual adjustments made by members of this elite to one another. No one else would have political rights or would be involved in either political or scientific persuasion about public affairs of any kind. These special people then, like Plato's philosopher-kings, would have to be insulated from other, merely private, interests. Since the public's consent to the favored political position of such philosopher-kings would of course have to be renewed daily, the rulers would have to engage in political persuasion in order to remain in power. So it would seem that *political* persuasion is the type of mutual adjustment essential to the achievement of a spontaneous order for any society that is not run, quite obviously, by a tyrant.

Since political persuasion can never operate in the manner of scientific persuasion, with truth as its highest standard incentive, political persuasion will make use of tricks, deals, compromises, and various forms of connivance, and each of the competing interest groups will always be trying to secure as much power for itself as it can. It must do so, in fact, in order to protect its interests. However, as we saw when we discussed Utilitarian views of freedom, such circumstances can result in the disintegration and destruction of a society.

A political community must therefore depend upon a fortuitous set of institutions to keep its political factions from destroying one another (and the state) in their efforts to achieve what power they can. Such institutions as these are necessary whether or not there is a traditional public respect for the ideal ends of the free associations we have been discussing. These institutions are largely fortuitous, since they could hardly be designed by any interest groups, unless by a group of interests *each one* of which was so unsure of being able to run things in its own interest that it must strive for a set of institutions that could not be captured, *in toto*, by any one existing faction.

Should such a set of institutions develop, we would then have a very good example of a higher-level moral sphere existing on the basis of a lower-level sphere of profit, power, and parochialism of interest—as both Madison and Hamilton clearly expected would be the case for the proposed United States. In a situation like that described, certain boundary conditions, supplied mainly by a set of institutions, would result in crasser interests being transformed, in operation, into moral principles, such as "justice."

Yet we must remember that the so-called political operations— those involved in securing office or influencing legislation—might remain very *un*moral indeed in a society like this. Nevertheless, the whole—if it did secure a distribution of power adequate to keep the different interests alive, and if the tradition of honoring the ideal objects of the various systems of spontaneous order remained sufficiently well established to leave these systems alone, to grow by the free mutual adjustments of their parts—if the whole should work in this way, it would be operating as well as could possibly be expected.

However, we will perceive that it is operating well only if we manage to abandon our deeply ingrained moral perfectionism—one of the causes, we may recall, of moral inversion. We could do this, of course, if we were to dwell fully in the view developed here, namely, that the moral level exists on the foundation of a lower, essentially nonmoral, level and that the latter must inevitably place limits on the accomplishments of the higher level. It is not difficult to see the conceptual meaning and ontological validity of this principle, but it may be difficult for us to live with it.

Once we have fully grasped the import of the necessary limits on our ability to construct a perfect society and can dwell in that import, we will refrain from various sorts of radical actions aiming at the full establishment of justice and brotherhood. We will recognize that we *can* reduce unjust privileges, but only by graded stages and never completely. No single panacea for them exists. They can be dealt with only one at a time, never wholesale, since we have to use the power of the present system in order to make any changes in it. To try to reform all the power structures at once would leave us with *no* power structure to use in our project. In any case, we will be able to see that *absolute* moral renewal could be attempted only by an absolute power and that a tyrannous force such as this must destroy the whole moral life of man, not renew it.

We will also acknowledge that the free and liberal society we are espousing is a conservative one as well. Our insistence upon holding that the independence of thought involved in science, art, and the legal profession is an inviolable right is really a subscription to a kind of orthodoxy which, although it does not specify a list of fixed articles of faith, is regarded as unchangeable. We will not, in other words, understand our society to be "open" in the direction of squashing these independent enclaves of thought and action; we will also freely admit that this orthodoxy is not only backed by the coercive power of the state but is financed by the beneficiaries of office and property— together with all the injustices and imperfections such arrangements must bring. We will perceive, therefore, that we have a moral allegiance to a manifestly imperfect, if not *immoral* society; and we will find, paradoxically, that our duty lies in the service of ideals which we nevertheless know we cannot possibly achieve.

It is at this point, however, that we may have some hesitation. As men we appear, unlike all other animals, to need a purpose that bears on eternity. Truth certainly does this, and so do our other ideals. We have called into existence the whole firmament of values that make up the cultural level of the life in which we dwell. These works of the imagination are like a tremendous burst of glory. Yet they are rooted ultimately in the lower physiological level of man, where he is a mere animal among animals. Indeed much of this burst of glory is due to the purely physiological structures that make speech possible for us. But these works of the imagination ascend from these lowly structures, as we have seen in this book, to levels of autonomous meaning of ever greater comprehensiveness—and we do not see the end in sight.

There is another level of meaningful attainment, however, mediating between these higher levels of man's culture and his lower physiological level. This is the level of his political associations. If men create a boundary for their animal lives that exhibits moral and political principles of the sort we have been outlining in this chapter, then all the other meaning levels we have been talking about in this book—science, art, religion, and the like—become possibilities. If men do not create or maintain this sort of boundary, these other higher levels must be hampered in their development—or die. However, as is true of the other works of man's imagination, the moral and political realms have principles of their own; and, if these are projected far beyond their essential connections with the lower ones of power and profit, in which they are rooted, they lead us to form a mental image—a utopian dream—of a world operating only in terms of these higher principles: just and pure and free from any contamination by the crasser elements from which they have sprung. The moral perfection of such a world beckons to us and makes us dissatisfied with our own moral shortcomings and those of our societies.

Perhaps it has been the clear moral call of Christianity that has left behind in us a distillation which causes us to burn with such a hunger and thirst after righteousness. If so, it should be possible for us to find in this same Christianity the antidote for this poison of moral perfectionism; for what this religion has also told us is that we are inescapably *im*perfect and that it is only by faith and trust in the all-encompassing grace of God that we can project ourselves into that supreme work of the imagination—the Kingdom of God—where we can dwell in the peace and hope of the perfection that is God's alone and thus where we can, in a wholly inexplicable and transnatural way, find our hunger and thirst after righteousness satisfied at last—in the midst of all our imperfections. As Saint Paul tells us his God told him: "I will not remove your infirmity. For my strength is made perfect in weakness."

Those of us who cannot in this way, through religion, sublimate our dissatisfaction with our own moral shortcomings—and with those of our societies—have a more difficult job. We must learn to suffer patiently the anguish these imperfect fulfillments cause us. A steady recognition that the evils which prevent the fullness of moral development are precisely the elements which are also the source of the power that gives existence to whatever moral accomplishments we see about us may eventually lead us to a tolerance of these lower elements similar to the tolerance we grant to the internal-combustion engine: it is noisy and smelly, and occasionally it refuses to start, but it is what gets us to wherever it is we get.

We must somehow learn to understand and so to tolerate—not destroy—the free society. It is the only political engine yet devised that frees us to move in the direction of continually richer and fuller meanings, i.e., to expand limitlessly the firmament of values under which we dwell and which alone makes the brief span of our mortal existence truly meaningful for us through our pursuit of all those things that bear upon eternity.

NOTES

Notes to Chapter One

1. B. F. Skinner, *Beyond Freedom and Dignity* (New York: Alfred A. Knopf, Inc., 1971).

2. Baron d'Holbach, *The System of Nature*, trans. H. D. Robinson (Boston: J. P. Mendum, 1853), pp. 153, ix-x.

3. W. E. H. Lecky, *History of the Rise and Influence of the Spirit of Rationalism in Europe*, 2 vols. (New York: Appleton, 1878), 1:128.

4. Hermann Rauschning, *The Revolution of Nihilism*, trans. Ernest W. Dickes (New York: Longmans, Green, 1939).

5. V. I. Lenin, "Where to Begin?" (1901) and "What Is to Be Done?" (1902) in *Collected Works*, ed. Victor Jerome, trans. Joe Fineberg and George Hanna (Moscow: Foreign Language Publishing House, 1961), 5:23-24, 473-84, and 514-18.

6. Konrad Heiden, *Der Fuehrer*, trans. Ralph Manheim (Boston: Houghton Mifflin, 1944), pp. 145-50.

Notes to Chapter Two

1. D. O. Hebb, "The Problem of Consciousness and Introspection," *Brain Mechanisms and Consciousness*, ed. J. F. Delafresnaye (Oxford: Basil Blackwell, 1954), p. 404; L. S. Kubie, "Psychiatric and Psychoanalytic Considerations of the Problem of Consciousness," ibid., p. 446; K. S. Lashley, "Dynamic Processes in Perception," ibid., p. 424.

2. Clyde Kluckhohn and Dorothea Leighton, *The Navaho* (Cambridge, Mass.: Harvard University Press, 1960), p. 177.

3. Gordon Childe, *What Happened in History* (Baltimore: Penguin Books, 1961), p. 16.

4. R. Pipes, "Russia's Intellectuals," *Encounter* 22 (1964): 79-84.

5. C. S. Lewis, *The Abolition of Man* (New York: Macmillan, 1947), pp. 28-33.

6. Leslie Paul, *The Annihilation of Man* (New York: Harcourt, Brace, 1945), p. 143.

Notes to Chapter Three

1. Gilbert Ryle, *The Concept of Mind* (London: Hutchinson, 1949), pp. 61, 23, 46, 58.

2. M. Merleau-Ponty, *Phenomenology of Perception*, trans. Colin Smith (London: Routledge, 1962), p. 185.

3. See, for example, B. F. Skinner, "Behaviorism at Fifty," *Behaviorism and Phenomenology*, ed. T. W. Wann (Chicago: University of Chicago Press, 1964), pp. 82-94.

4. M. Polanyi, *Personal Knowledge* (Chicago: University of Chicago Press, 1958; rev. ed., New York: Harper Torchbooks, 1964), pp. 370-71.

5. Noam Chomsky, review of B. F. Skinner, *Verbal Behavior, Language* 35 (1959): 26-58.

6. My theory of irreducible levels goes back to my *Personal Knowledge* and has since been developed further in a number of stages, a survey of which is to be found in my book *The Tacit Dimension* (Garden City, N.Y.: Doubleday, 1966). See also my paper "Life Transcending Physics and Chemistry" (1967). An application of the theory to the body-mind relation is to be found in my paper "The Structure of Consciousness" (1965). Both these papers have been republished in *Knowing and Being*, ed. Marjorie Grene (Chicago: University of Chicago Press, 1969).

F. S. Rothschild anticipated my conclusion that the mind is the meaning of the body. His writings extend back to 1930. They are summarized in *Das Zentralnervensystem als Symbol des Erlebens* (Basel and New York: K. Karger, 1958), with a briefer summary in English in 1962: "Laws of Symbolic Mediation in the Dynamics of Self and Personality," *Annals of the New York Academy of Sciences* 96 (1962): 774-84. He has developed this idea widely in neurophysiology and psychiatry, where I am not competent to follow him.

7. I. Kant, *Critique of Pure Reason*, trans. Norman Kemp Smith (New York: St. Martin's Press, 1929), p. 183 (p. A141).

8. F. Waismann, "Verifiability," *Proceedings Aristotelian Society*, supp. 19 (1945): 121-33.

9. In his contribution to the Symposium on Logic and Psychology at the Convention of the American Psychological Association in 1967, Quine ascribed the identification of specimens of a class to undefined native powers and also showed that the general features of a class cannot be empirically demonstrated. He regards this as an instance of the inscrutability of the categoricals. See also W. V. O. Quine, *Word and Object* (Cambridge and New York: The Technology Press of The Massachusetts Institute of Technology and John Wiley & Sons, Inc., 1960), pp. 60-80.

10. J. S. Mill, *A System of Logic, Ratiocinative and Inductive* (New York:

Harper & Bros., 1900), pp. 332-72. C. G. Hempel, *Scientific Explanation*, Forum Philosophy of Science Series (Washington, D.C.: U.S. Information Agency, 1964). C. G. Hempel and P. Oppenheim, "Studies in the Logic of Explanation," *Philosophy of Science* 15 (1948): 135-75. E. Nagel, *The Structure of Science* (New York: Harcourt, Brace & World, 1961).

11. M. Scriven, "Explanation, Prediction and Laws," *Minnesota Studies in the Philosophy of Science*, ed. Herbert Feigl et al. (Minneapolis: University of Minnesota Press, 1962), 3:172.

12. M. Polanyi, *Science, Faith and Society* (London: Oxford University Press, 1946; reprinted, Chicago: University of Chicago Press, 1964), p. 24.

13. H. Poincaré, *Science et méthode* (Paris: Flammarion, 1908). J. Hadamard, *The Psychology of Invention in the Mathematical Field* (Princeton: Princeton University Press, 1945). G. Polya, *How to Solve It* (Princeton: Princeton University Press, 1945).

14. J. B. Conant, *Harvard Case Histories in Experimental Science* (Cambridge: Harvard University Press, 1957). T. S. Kuhn, *The Structure of Scientific Revolutions* (Chicago: University of Chicago Press, 1962). L. K. Nash, *The Nature of the Natural Sciences* (Boston: Little, Brown, 1963). W. Whewell, *Philosophy and Discovery* (London: John W. Parker, 1890).

15. *Science et méthode*, pp. 50-63.

Notes to Chapter Four

1. Edward Sapir, *Culture, Language, and Personality* (Berkeley: University of California Press, 1956), pp. 8-15. Bertrand Russell, *An Inquiry Into Meaning and Truth* (New York: W. W. Norton & Co., 1940), p. 84. Susanne K. Langer, *Philosophy in a New Key* (Cambridge: Harvard University Press, 1942), pp. 58-59. Erwin W. Strauss, *Phenomenological Psychology* (New York: Basic Books, 1966), pp. 183-84.

2. Aristotle, *Basic Works*, ed. Richard P. McKeon, trans. Ingram Bywater (New York: Random House, 1941), pp. 1479 (1459a), 1476 (1457b).

3. Owen Barfield, "Poetic Diction and Legal Fiction," in *Essays Presented to Charles Williams* (Oxford: Oxford University Press, 1947), p. 111.

4. *Basic Works*, p. 1479 (1459a).

5. I. A. Richards, *The Philosophy of Rhetoric* (New York and London: Oxford University Press, 1936), p. 127.

6. Max Black, *Models and Metaphors* (Ithaca, N.Y.: Cornell University Press, 1962), pp. 45, 38, 39, 46.

7. André Breton, *Les Vases communicants* (Paris: Gallimard, 1955), p. 148. English translation from Richards, *The Philosophy of Rhetoric*, p. 123.

8. Ezra Pound, "In a Station in the Metro," *Selected Poems*, ed. T. S. Eliot (London: Faber & Gwyer, 1928), p. 89.

9. W. B. Yeats, "Three Movements," *The Collected Poems of W. B. Yeats* (New York: Macmillan, 1956), p. 236.

10. T. S. Eliot, "East Coker," Part IV, *The Complete Poems and Plays* (New York: Harcourt, Brace, 1952), p. 127.

11. William Shakespeare, *Richard II*, act 3, scene 2.

12. Hidden within this metaphor there is, of course, another metaphor— the metaphorical use of an ointment in the performance of what we call a rite or a ritual. (In a later chapter we will be discussing the sort of meaning that rituals achieve.) The king's irrevocable power has been conferred on him by a rite involving the use of a balm. This notion of a ritual ordination operates as a subsidiary clue (namely, a tacit understanding of the divine right of the king to his office) in the total meaning that the imagination puts into the metaphor; but we should not allow this feature of this metaphor to obscure for us the literal, *naturalistic* meaning that the words themselves have. Otherwise the metaphor cannot make its point. The words presented to us treat the balm as a fluid of some sort that cannot be washed away by all the quantity and power of the seas. Period. This asserts a preposterous (but striking) natural relation between these two fluids which, *as such*, is void of any metaphorical meaning the balm has as part of a ritual. Such a naturalistically absurd statement is essential to the type of meaning that our imagination creates in and through a metaphor.

13. William Empson, *The Structure of Complex Words* (London: Chatto & Windus, 1951), pp. 346–49.

14. Charles Baudelaire, "Les Fleurs du Mal," *Oeuvres Complètes de Charles Baudelaire*, ed. Jacques Crépet, 19 vols. (Paris: Louis Conard, 1930), 1:15.

15. Dom Moraes, *My Son's Father* (London: Secker & Warburg, 1968), p. 191.

16. I. A. Richards, *Principles of Literary Criticism* (New York: Harcourt, Brace, 1942), pp. 145–46.

Notes to Chapter Five

1. I. A. Richards, *Principles of Literary Criticism* (New York: Harcourt, Brace, 1942), p. 110.

2. Ibid., p. 145.

3. Susanne Langer also makes this point, showing us that detachment and otherness are essential to works of art. See her *Feeling and Form* (London: Routledge & Kegan Paul Ltd., 1953), pp. 45, 46.

4. Immanuel Kant, *Critique of Aesthetic Judgement*, trans. James C. Meredith (Oxford: Clarendon Press, 1911), p. 49. Conrad Fiedler, *On Judging Works of Visual Art* (Berkeley and Los Angeles: University of California Press, 1949), p. 60; see also pp. 52, 57.

5. *Principles of Literary Criticism*, p. 237.

6. T. S. Eliot, "The Metaphysical Poets," *Selected Essays* (New York: Harcourt, Brace, 1932), p. 247.

7. E. H. Gombrich, *Art and Illusion* (London: Phaidon Press, 1962), p. 236. (The date of Quartremère de Quincy's work is 1823.)

8. Irving Rock and Charles S. Harris, "Vision and Touch," *Scientific American* 216 (May 1967): 96–104. H. Kottenhoff, L. E. H. Lindal, and S. E. R. Mable, "Optical and Mechanical Devices for Testing Susceptibility to Motion Sickness," *Perceptual and Motor Skills* 7 (1957): 221–22.

9. M. H. Pirenne, "Les Lois de l'optique et la liberté de l'artiste," *Journal de psychologie normale et pathologique* 60 (1963): 151–66.

Notes to Chapter Six

1. H. W. Janson, *History of Art* (New York: Harry N. Abrams, 1962), p. 11.

2. Ibid., p. 10.

3. Pierre Laplace, *A Philosophical Essay on Probabilities*, trans. F. W. Truscott and F. L. Emory (New York: John Wiley & Sons, 1902), p. 4.

Notes to Chapter Seven

1. E. H. Gombrich, *Art and Illusion* (London: Phaidon Press, 1962), pp. 85–86, 91–101.

2. Charles Baudelaire, *Oeuvres complètes de Charles Baudelaire*, ed. Jacques Crépet, 19 vols. (Paris: Louis Conard, 1930), 1:7.

3. Arthur Rimbaud, "Bateau ivre," *Oeuvres de Arthur Rimbaud* (Paris: Mercure de France, 1937), p. 85. The translation in the text is by M. P.

4. Arthur Rimbaud, Letter to Paul Demeny, 15 May 1871, in *Complete Works and Selected Letters*, ed. and trans. Wallace Fowlie (Chicago: University of Chicago Press, 1966), p. 307.

5. Leo Tolstoy, *What Is Art?*, trans. Aylmer Maude (Indianapolis: Library of Liberal Arts, 1960), pp. 76–99 (chap. 10).

6. I. A. Richards, *Principles of Literary Criticism* (New York: Harcourt, Brace, 1942), p. 293.

7. Alain Robbe-Grillet, *For a New Novel*, trans. Richard Howard (New York: Grove Press, Inc., 1965), pp. 154–56.

8. Helmut Kuhn, *Wesen und Wirken des Kuntswerks* (Munich: Kösel-Verlag, 1960), pp. 67, 68. Translated by M. P.

9. Ibid., pp. 67–73. See also *Schriften zur Ästhetik* (Munich: Kösel-Verlag, 1966), p. 271.

Notes to Chapter Eight

1. Erwin W. Strauss, *Phenomenological Psychology* (New York: Basic Books, 1966), pp. 144, 142.

2. Mircea Eliade, *Myth and Reality: World Perspectives*, trans. Willard R. Trask (New York: Harper & Row, 1963), pp. 5–6.

3. Ibid., p. 11.

4. Mircea Eliade, *Images and Symbols*, trans. Philip Mairet (London: Harvill Press, 1961), p. 57.

5. *Myth and Reality*, p. 140; see also p. 139.

6. Ibid., p. 19.

7. Ibid., p. 18.

8. Ernst Cassirer, *The Philosophy of Symbolic Forms*, trans. Ralph Manheim, 3 vols. (New Haven: Yale University Press, 1955), 2:77-78.

9. *Myth and Reality*, p. 19.

10. *Images and Symbols*, p. 59.

11. Ibid., p. 12.

12. *Myth and Reality*, pp. 32-33.

13. Ibid., p. 143.

14. Toshimitsu Hasumi, *Zen in Japanese Art*, trans. John Petrie (London: Routledge, 1962), p. x.

15. Mircea Eliade, *The Two and the One*, trans. J. M. Cohen (London: Harvill Press, 1965), pp. 88-89.

16. Ibid., pp. 80-81. Eliade refers to the use by C. G. Jung of the term *coincidentia oppositorum* as essential to a description of the process of individuation and of the ultimate aim of the whole psychic activity. What Eliade referred to then in Jung may now be found in C. G. Jung, *Psychology of the Transference, The Collected Works of C. G. Jung*, ed. Herbert Read, Michael Fordham, and Gerald Adler, trans. R. F. C. Hull, 17 vols. (New York: Pantheon Books, 1954), 16:163-321, and in *Mysterium Coniunctionis, Collected Works* (Princeton: Princeton University Press, 1970), vol. 14.

17. Hasumi, *Zen in Japanese Art*, p. 14.

18. Mircea Eliade, *Myths, Dreams and Mysteries*, trans. Philip Mairet (New York: Harper & Bros., 1960), pp. 195-96.

Notes to Chapter Nine

1. Lucien Lévy-Bruhl, *How Natives Think*, trans. Lilian A. Clare (London: Allen & Unwin, 1926), pp. 352-58.

2. Claude Lévi-Strauss, *The Savage Mind* (London: Weidenfeld & Nicolson, and Chicago: University of Chicago Press, 1966), pp. 267-69.

3. Immanuel Velikovsky, *Worlds in Collision* (New York: Macmillan, 1950).

4. Ralph E. Juergens, "Minds in Chaos: A Recital of the Velikovsky Story," *American Behavioral Scientist* 7 (1963): 4-17. Livio C. Stecchini, "The Inconstant Heavens: Velikovsky in Relation to Some Past Cosmic Perplexities," ibid., pp. 19-44. Alfred de Grazia, "The Scientific Reception System and Dr. Velikovsky," ibid., pp. 45-49, and *The Velikovsky Affair: The Warfare of Science and Scientism* (New York: New York University Books, 1966).

5. Michael Polanyi, "The Growth of Science in Society," *Minerva* 4 (summer 1967): 533-45, and *Criteria for Scientific Development, Public*

Policy, and National Goals, ed. Edward Shils (Cambridge, Mass.: Massachusetts Institute of Technology Press, 1968), pp. 187-99.

6. Ernst Cassirer, *The Philosophy of Symbolic Forms*, trans. Ralph Manheim, 3 vols. (New Haven: Yale University Press, 1955), 2:40-42.

7. Ibid., pp. 49-51.

8. See Lévy-Bruhl, *How Natives Think*, p. 77.

9. Ibid., pp. 78-79.

10. Mircea Eliade, *Images and Symbols*, trans. Philip Mairet (London: Harvill Press, 1961), p. 59:

Notes to Chapter Ten

1. It is interesting to note in this connection how John Dewey distinguished his views from those of "militant atheism." He pointed to the lack of "natural piety" in such atheism, describing it as the attitude of one who thinks of himself as an "isolated and lonely soul . . . living in an indifferent and hostile world and issuing blasts of defiance." Dewey protested, however, that nature "produces [not only] what occasions discord and confusion [but also] whatever gives reinforcement and direction." He obviously believed, therefore, that nature not only bequeaths us our problems but also supplies us with resources for their solutions. This conviction is, of course, essentially a "faith," since no one can *know* that a problem not yet solved can in fact *be* solved. But it is also an essential faith if one is ever to find solutions to problems. We can find only by searching, and we can search only if we believe there is something to be found—which is believing that we do *not* live in a "hostile" universe but rather in one upon which we can rely for support. And this, Dewey claimed, is essentially "a religious attitude." (See John Dewey, *A Common Faith* [New Haven: Yale University Press, 1934], pp. 52-54.)

2. Aristotle, *Basic Works*, ed. Richard P. McKeon, trans. Ingram Bywater (New York: Random House, 1941), p. 1482 (1460a).

3. William James, "The Will to Believe," in *The Will to Believe and Other Essays* (New York: Longmans, Green, 1898), p. 25.

4. Jean-Paul Sartre, *Existentialism*, trans. Bernard Frechtman (New York: Philosophical Library, 1947), p. 26.

Notes to Chapter Eleven

1. Charles Saunders Peirce, *Collected Papers*, ed. Charles Hartshorne and Paul Weiss, 8 vols. (Cambridge, Mass.: Harvard University Press, 1960), 7:15-20 (6.13-24), 25-27 (6.30-34), 43-44 (6.62-64), 132-40 (6.185-206). William James, *A Pluralistic Universe* (New York: Longmans, Green, 1920), pp. 30-34, 303-8, 321-28.

2. Alfred North Whitehead, *Adventures of Ideas* (New York: The Free Press, 1967), pp. 129-30, 146-50.

3. See, for instance, C. F. A. Pantin, *The Relation between the Sciences*

(Cambridge, Eng.: At the University Press, 1968), pp. 35-45, 53.

4. G. W. von Leibniz, "The Principles of Nature and Grace," pp. 1039-40 in *Philosophical Papers and Letters*, trans. and ed. Leroy E. Loemker (Chicago: University of Chicago Press, 1956).

5. H. Spemann, *Embryonic Development and Induction* (New Haven: Yale University Press, 1938). Paul Weiss, *Principles of Development* (New York: Henry Holt, 1939). C. H. Waddington, *The Strategy of the Genes* (London: Allen & Unwin, 1957), and *New Patterns in Genetics and Development* (New York: Columbia University Press, 1962).

6. See, for instance, a recent book edited by Howard H. Pattee, *Hierarchy Theory* (New York: George Braziller, 1973).

7. George Santayana, *Reason in Religion* (New York: Charles Scribner's Sons, 1948), p. 5.

8. George Santayana, *Obiter Scripta*, ed. Justus Buchler and Benjamin Schwarth (New York: Charles Scribner's Sons, 1936), p. 296.

9. James, *The Will to Believe and Other Essays*, pp. 25-27.

Notes to Chapter Twelve

1. See Michael Polanyi, *Science, Faith, and Society* (Chicago: University of Chicago Press, 1964; first published, 1946), for an early statement about the traditional grounds of science and the cultivation of originality. These ideas were partly developed earlier than this, in essays written in the 1940s and published in *The Logic of Liberty* (Chicago: University of Chicago Press, 1951), and they later formed the basis for *Personal Knowledge* (Chicago: University of Chicago Press, 1958). More recent statements of these matters may be found in "Science: Academic and Industrial," *Journal of the Institute of Metals* 89 (1961): 401-6; in "The Republic of Science," "The Potential Theory of Adsorption," and "The Growth of Science in Society," reprinted in *Knowing and Being*, ed. Marjorie Grene (Chicago: University of Chicago Press, 1969); and in the book *The Tacit Dimension* (Garden City, N.Y.: Doubleday, 1966).

2. This view has been persuasively expressed by K. R. Popper, for example in his *Logic of Scientific Discovery* (New York: Basic Books, 1959), p. 279, as follows:

"But these marvellously imaginative and bold conjectures or 'anticipations' of ours are carefully and soberly controlled by systematic tests. Our method of research is not to defend them, in order to prove how right we were. On the contrary, we try to overthrow them. Using all the weapons of our logical, mathematical, and technical armoury we try to prove that our anticipations were false—in order to put forward, in their stead, new unjustified and unjustifiable anticipations, new 'rash and premature prejudices,' as Bacon derisively called them."

Popper even goes so far as to hold (on p. 419) that scientists will choose the *least* probable hypothesis to investigate, since it will be the easiest one to refute! Such a supposition borders on the ridiculous, even though "improbable" does not mean "implausible" to him. Scientists do, he thinks, choose the most general *plausible* hypothesis to investigate. But, in doing so, they in fact choose the least probable one. He thinks the most general plausible hypothesis is the least probable one because, being the most general, it will be the one farthest removed from sense data that could count as evidence for or against it. But, having noted this merely coincidental circumstance about the most general plausible hypothesis, Popper then goes on to suppose that scientists do not choose it simply because it is the most *general* plausible one they can think of, but rather because it is the most *risky* one they can think of. The cart has not only been placed before the horse, but the horse has also been turned completely around.

Notes to Chapter Thirteen

1. James Madison, "Federalist Paper No. 10," *Selections from the Federalist Papers*, ed. Henry S. Commager (New York: Appleton-Century-Crofts, 1949), p. 14.

2. Adam Smith, *The Wealth of Nations* (New York: Modern Library, 1937), pp. 248-50, 734-40.

3. It should be noted, however, that markets, pricing, interest rates, profits, etc., are made necessary, not by *private capitalism*, but by the existence of an *industrial mode of production*, which otherwise has no way of adjusting its many separate parts to one another in a meaningful way, since it has no other way of evaluating these *in terms of one another* and of the actual wants of actual consumers. These various so-called capitalistic institutions are as essential in a socialist society, therefore, as in a capitalistic one. A centrally planned modern economy is a physical impossibility. See Michael Polanyi, *The Logic of Liberty* (Chicago: University of Chicago Press, 1951), pp. 111-53. See also Paul Craig Roberts, *Alienation and the Soviet Economy* (Albuquerque: University of New Mexico Press, 1971), pp. 48-88.

4. It is an interesting irony, however, that a privileged class seems to make its appearance also in noncapitalist societies—although here it becomes merely a caricature of its capitalist counterpart. See Milovan Djilas, *The New Class* (New York: Frederick A. Praeger, 1957), pp. 54-56.

BIBLIOGRAPHICAL NOTE

The chapters in this book are adapted by the authors from the following published and unpublished works of Michael Polanyi.

Chapter 1.   "Sixty Years in Universities." Unpublished lecture given at the University of Toronto, 24 November 1967.
The Logic of Liberty. Chicago: University of Chicago Press, 1951. Chapter 7.

Chapter 2.   "Sixty Years in Universities."
"On the Modern Mind." *Encounter* 24 (May 1965): 12-20.
"The Study of Man." *Quest* (Bombay) no. 29 (April-June 1961): 26-35.
"Logic and Psychology." *American Psychologist* 23 (January 1968): 27-43.

Chapter 3.   "Logic and Psychology."
"On the Modern Mind."

Chapter 4.   "From Perception to Metaphor." Unpublished lecture given at the University of Texas and the University of Chicago, February and May 1969.
"Meaning." Unpublished lecture given at the University of Chicago, April 1970.

Chapter 5.   "Works of Art." Unpublished lecture given at the University of Texas and the University of Chicago, February and May, 1969.
"Meaning."

Chapter 6.   "Works of Art."
              "Meaning."

Chapter 7.   "Visionary Art." Unpublished lecture given at the
              University of Texas and the University of Chicago,
              February and May 1969.

Chapter 8.   "Visionary Art."
              "Myths, Ancient and Modern." Unpublished lecture
              given at the University of Texas and the University of
              Chicago, February and May 1969.

Chapter 9.   "Myths, Ancient and Modern."

Chapter 10.   "Acceptance of Religion." Unpublished supplement to
              the lectures given at the University of Texas and the
              University of Chicago, February and May 1969.
              "Meaning."

Chapter 11.   "Expanding the Range." Unpublished lecture given at
              the University of Texas, April 1971.

Chapter 12.   "Honor." Unpublished lecture given at the University of
              Texas, April 1971.
              The Tacit Dimension. Garden City, N.Y.: Doubleday &
              Co., Inc., 1966. Pages 63–79.

Chapter 13.   *The Logic of Liberty*. Chapters 3, 8, 9, and 10. *Personal
              Knowledge*. Chicago: University of Chicago Press,
              1958. Chapter 7.

INDEX

Absurdity, 25, 26, 77, 78, 112, 113, 115, 134, 139-40, 143, 144, 145, 161, 163, 179, 181

Acceptance: elements in, 149-51; kind of knowledge related to, 149; of a scientific claim, 195

Achievement: category of, required in living things, 170; entails possibility of failure, 141, 170; excitement of, 196

Adjustments, mutual, 190, 198-200, 203, 204, 207, 208, 210-13; interference with, sometimes required, 204. *See also* Freedom of mutual adjustment

Adjustments, reflexive, 207

Aesthetics, 95; cornerstone of, 106; visionary, 129; in Zen Buddhism, 128-29

Age of Reason, 10

Alter ego, principle of, 136

Ambition, 196, 208

Ames, A., 42

Analysis, alternates with integration, 141

Anaxagoras, 161

Anglo-American liberalism: origins of, 6; religious character of, 11; restraints upon, 11; suspended logic of, 10, 14

Anthropologists, 26

Anticipation of a hidden truth, 193. *See also* Intuition, questing

Antinovel, 110, 114

Aquinas, Saint Thomas, 19

Archaic people, 135, 145. *See also* Mind, archaic; Thought, archaic

Aristotle: *Poetics*, 75, 158; struggle against the authority of, 7

Aristotelian cosmology, 142, 174

Arrogance, 157

Art: concrete definition of, 204; as creative growth of man's existence, 99, 109 (*see also* Meaning, addition of, by art; Integration of our diffuse experiences; Integration, self-giving); as imitative, 109-10, 150; immortality of, 109; as interaction of means and ends, 99; modern, contributed to destruction of coherence, 116; modern, and Zen Buddhism, 129-30; no external tests of, 103; public now able to grasp imaginative meaning of, 93-94; representative, 83-92, 149-50; if representative, "story" must have plausibility, 158; in Zen Buddhism, 129

Art critics: bohemian moderns, 93; bourgeois academics, 93

Art, work of: acceptance of, 149; composed of frame and story, 86, 88, 90-91, 97-98; detachment of, 102, 109, 124-25, 150; grasped only by imagination, 94; as outside time, 118 (*see also* Artist, sees subject as in one moment); as production of something only vaguely anticipated, 98; repre-